Weingart's Dictionary and Manual of Fireworks

ISBN: 978-0-8206-0093-2

Chemical Publishing Company:
www.chemical-publishing.com
www.chemicalpublishing.net

Printed in the United States of America

DICTIONARY and MANUAL
of
PYROTECHNY.

Covering the authors work and experiments
from 1890 to 1935.

A drop of ink, t'is said
made millions think:

A spark of fireworks has oft'
made thousands blink.

Though handicapped in the struggle of life, I have achieved some small success through the constant untiring assistance of my wife. As a slight token of appreciation I dedicate to her this work.

The Author.

Preface.

In following work the object has been to place on file a complete record of the developements of the art of pyrotechny up to the present time. Previous epochs, in what may be considered the modern manufacture of fireworks, have been punctuated by Ruggiere, 1812 in Italian; Chertier, 1854 in French; Kentish, 1878 in English and Antoni 1893 in Italian.

There have been writings on the subject as early as the year 1225 and the first authentic manual seems to be in German, written in 1432 but these are mainly museum curiosities. Since 1893 some historical and specialized articles have appeared in connection with pyrotechny but nothing of a general or comprehensive character in the way of a working manual.

In the interim great strides have been made, both in the materials used as well as in the methods employed in the mass production of articles in general use for public celebrations.

Powdered aluminum has added many beautiful effects; picric acid is extensively used and potassium perchlorate has greatly reduced the danger from spontaneous combustion. Phosphorus, both yellow and amorphous is largely used and machines to replace manual operations have been devised in many instances.

The formulars herein given are all taken from those in actual use and will positively produce the effects for which they are indicated. The machines and tools shown are those in use in the largest and most recently equipped factories.

The article on Chinese Firecrackers is the first detailed description of this interesting article ever written in the English language.

A caution added here might not be amiss. Never hold any kind of fireworks in the hand while it is burning. Make this rule without exception because sometimes the most unexpected explosions occur and cause painful injury. If firing shell be sure never to look into the mortar after the shell has been inserted. A spark might have in some manner gotten inside. Also never allow the arm or hand to come over the muzzle of the mortar. Long sections of bare match should be used on shells so that the one lighting them has ample time to reach a safe distance in case of the bursting of the gun. There is always enough danger to avoid all unnecessary risk.

The beginner is advised to read carefully and memorize the instructions given under "manipulations" before attempting the production of any kind of fireworks. Actions which seem trivials are sometimes fraught with gravest danger. The object of fireworks is to give pleasure so do not compound it with tragedy.

In conclusion I might add that in this work pyrotechny is treated rather as a craft than as an art. For the artistic feature, imagination only is requisite but to work out the results of imagination is just plain tedious, patient and often hard work. Consequently, if we have the methods at hand it will be easier to bring to life the artistic ideas.

January 30th. 1937.

INTRODUCTION.

To those contemplating the making of fireworks, either professionally or as an amusement it is desirable to understand the principles which govern the operations of the various devices as well as the compositions of the chemicals entering into their production.

The principle of colored lights is based on:

1st. Producing a mixture that will burn at a reasonable speed while generating an intense heat,

2nd. Adding thereto the salts of such elements, in the spectrum of which, predominate lines of the desired colors.

Heat generating compounds consist chiefly of:

(a). Substances which yield oxygen freely when ignited in the presence of carbon, viz:

Potassium chlorate,
Potassium per-chlorate,
Potassium nitrate,
Sodium chlorate,
Barium nitrate,
Strontium nitrate.

(b). Carbon and carbonaceous sources, viz:

Charcoal,
Shellac,
Fossil gums,
Resins,
Asphaltum,
Dextrine,
Stearine,
Sugar of milk,
Corn flour etc.

In addition to the above there are some substances which when added to colored fire compositions increase the affinity of the several constituents for one another thereby improving the colors, viz:

Sulphur
Picric acid

and in the instance of blue and green fires it is almost essential to add an easily volatilized chloride in order to get sufficient depth of color.

viz: Calomel
Sal Ammoniac.

The exact function of these last named substances is not entirely clear but it appears that the best spectrums are yielded by the chlorides of the elements. However most chlorides are deliquescent and therefore unsuited to fireworks making. By adding a substance that yields chlorine freely at the moment of decomposition the necessary conditions are produced for obtaining the best results.

The following substances are most generally used for producing pyrotechnical colors.

PURPLE

Strontium and Copper compounds.

ORANGE

Strontium and Sodium salts.

Bright or "Plain" Mixings

Consist almost entirely of:

Saltpeter,
  rarely Lead nitrate,
Sulphur,
Charcoal or lampblack.

with the addition of

Steel filings,
Iron borings,
Zinc powder,
Antimony,
Orpiment,
Realgar.

PINK

Calcium carbonate,
Calcium sulphate,
Calcium oxalate.

GREEN

Barium nitrate,
Barium chlorate,
Boric acid,
and rarely Thalium nitrate.

RED

Strontium nitrate,
Strontium carbonate,
and rarely Lithium salts.

BLUE

Copper carbonate,
Copper arsenate,
Copper sulphate,
Copper black oxid,
Copper & Ammonium sulphate,
Copper oxalate,
Copper & Ammonium chloride.

YELLOW

Sodium oxalate,
Sodium bicarbonate,
Sodium metantimoniate.

The intensly bright white sparks are produced by aluminum powder. At one time Magnesium was used for this purpose but it has been entirely discontinued.

PART I.

INGREDIENTS

SALTPETER
(Nitre—Potassium nitrate)

This most important ingredient is produced in New York state of so high a quality that it is needless to look for better. The most suitable for pyrotechnical purposes generally is the "double refined, powdered" which can be obtained at from 5c to 15c lb. in barrels of 350 lbs. the price varying according to market conditions. For some large work granulated saltpeter is used, burning slower and being cheaper at the same time. The most suitable is "Dupont #2".

Specifications for saltpeter to be used in fireworks making call for a salt that is clean, white and should be ground fine enough to pass through a sieve of 80 to 100 mesh. It should contain less than 1% of sodium, calcium and magnesium salts combined.

POTASSIUM CHLORATE
$K\ Cl\ O_3$

This very necessary chemical is now also being prepared in this country, near Niagara Falls, of an excellent quality. The price varies considerably though it usually declines somewhat after July 4th. The powdered ranges from 9c to 16c lb. in kegs of 112 lbs. For pyrotechnical purposes it should be white, odorless and contain not over ½ of 1% of sodium, calcium and bromine combined. It should be of the same fineness as saltpeter.

POTASSIUM PERCHLORATE.

$K\ Cl\ O_4$

This recently produced substance forms another valuable addition to the pyrotechnists art. Containing even more oxygen than the chlorate it is less liable to decomposition due to the fact that it is a salt of perchloric acid which is a much more stable acid than chloric from which the chlorate is derived. It can be substituted for the chlorate in most mixings and can be safely used in connection with sulphur. The price is slightly higher than the chlorate and its specifications are practically the same

SULPHUR.

The "flour of sulphur" which is used almost exclusively is made of good quality in New York state from the product of the Louisiana mines and is sold at 2c to 3c lb. in barrels of 250 lbs. Italian washed sulphur is recomended by some of the older English pyrotechnists for use with chlorate of potash but the writer never uses formulas containing the two substances, if possible. It is almost white and comes in bags of 50 lbs. "Flowers" of sulphur is also sometimes used as well as coarsely ground sulphur which burns somewhat slower than the first two varieties. Specifications call for less than 1/10 of 1% of impurities and the finely ground should pass through a sieve of 120 mesh.

CHARCOAL.

Willow coal is the best for fireworks purposes through coal made from any soft wood is suitable. Pine coal is not very desirable. Excellent charcoal is made near Rochester, N. Y. and can be had finely powdered, granulated or mixed @ 1½¢ lb. in barrels or sacks. Charcoal that has a brown tint indicates incomplete carbonization and should be avoided. Also it should contain a minimum of grit. Shaking a sample in a bottle of water and decanting several times will disclose an excessive amount of sand etc.

LAMPBLACK.

Germantown lampblack is very popular with pyrotechnists though there are a number of good brands on the market. To make a good bright star it should be free from oil or other impurities and it is sometimes necessary to bake it as will be explained later, in order to get rid of volatile impurities which impair its briliancy in burning. It can be bought in barrels of 1 lb. packages @ 3¢ lb.

SHELLAC

and other gums etc.

Shellac, a gum like substance, is the secretion of an insect living on a large variety of trees in northern India. After going through various processes it finally reaches this country in some dozen different grades. The T N is a good grade for fireworks making and costs 25¢ to 35¢ lb. As it is practically impossible to powder this article oneself, a ball mill being necessary, great care must be excercised in purchasing powdered shellac as it is frequently adulterated with sand etc. For the best work shellac is almost indispensable but for stock goods, tableau fires and torches a number of substitute gums have been introduced such as Kauri, a fossil resin of a light yellow to a dark brown color, obtained from New Zeland. Red Gum comes from the Kangaroo Islands, 5 miles from Australia. K. D. dust is used for green fire. Asphaltum produces excellent colors when finely ground but owing to its containing sulphur, or perhaps on account of being so easily decomposed it is liable to cause spontaneous combustion when mixed with potassium chlorate. A mixture of these will explode violently when struck with a hammer on an anvil. With potassium perchlorate, however, it is entirely safe. The Syrian Asphaltum is the best. So called Green Gum is merely powdered coconut shells and has no more value in pyrotechny than sawdust. Flour, dextrine, sugar of milk etc. are also frequently used as sources of carbon. Another article of this character is

STEARINE.

In making blue fire it has been found that stearine produces a better effect, especially with paris green and other copper salts, than most any of the other hydrocarbons. It is mostly obtained in cakes and is reduced to a servicable condition by setting a carpenter's plane upsidedown over a box and shoving the cakes against the blade so as to shave the stearine as fine as possible. When it is then mixed with the other ingredients it will pass through an ordinary sieve.

STRONTIUM NITRATE.

This article is made principally in England. Its chief source is celestite which is also shipped to Germany and this country for conversion into the nitrate. It is put up in kegs of 110 lbs. and casks of 600 lbs. in a sufficiently pure condition for use as received, costing here from 6c to 8c lb. It is probably the most useful color producing chemical used in fireworks making, as the deep red light which it gives is the most marked effect which the pyrotechnist has achieved. Owing to its deliquescent properties, however, a number of methods have been devised to overcome this tendency one of which is to melt in an iron pot over a fire some shellac and stir in the nitrate of strontia, cooling and pulverizing. Another plan is to use carbonate of strontia but at the cost of considerable depth of color. Strontium nitrate is used in a somewhat coarser powder than the potassium salts but should be a clean white and contain not over 1/5 of 1% moisture and ¼ of 1% sodium salts.

STRONTIUM CARBONATE.

In damp climates there is no alternative but to use this strontium compound for most exhibition work as a

piece of lancework made with nitrate of strontium, if exposed for one hour to a damp atmosphere, will hardly burn. Precipitated carbonate of strontium is the only kind which should be used and may be purchased for about 16c lb. or can be easily made by adding carbonate of ammonia to a solution of strontium nitrate, thoroughly washing and drying the precipitate. If sodium carbonate is used as a precipitant it is almost impossible to remove every trace of it from the carbonate of strontium and causing an orange tint to the red light.

BARIUM NITRATE.

Like strontium this chemical also comes to us mainly from England and Germany in similar packings and costs usually from 5c to 7c lb. on this side. As a color producer it is far inferior to strontium though it does not attract moisture. If used without calomel its color is so pale as to be almost indistinguishable from white. Specifications for fireworks making are practically the same as for strontium nitrate. A better salt for making green fire is

BARIUM CHLORATE.

This salt give a very beautiful emerald color but its high cost, viz: about 30c lb. makes it little used except in exhibition work. Some recipies have been given for green fire using boracic acid, thalium salts etc. but if used at all it is to a very limited extent. All barium salts are very poisonous.

SODIUM OXALATE.

It is a strange fact that while yellow is the most common color of fires in general its practical production in pyrotechny is accompanied with some difficulty from the fact that there is practically only one insoluble salt of sodium while all the others are more or less hygroscopic. The nitrate and bicarbonate give deep yellow lights but the least dampness will render them incombustible and even the oxalate, will in damp weather, attract moisture. The exception is Sodium met-antimoniate but as this salt costs $4.00 lb. and at best gives a pale color, it is not much used. Oxalate of soda costs about 20c lb. or can be easily made by adding bicarbonate of soda to a hot concentrated solution of oxalic acid. A copius precipitate falls which however cannot be washed but must be dried on a filter. An excess of oxalic acid should be maintained in this operation.

COPPER ARSENITE

(Paris green)

This article is made in New York state and elsewhere in this country and can be bought for from 10c to 15c lb. from dealers in painters supplies. It can also be easily made by adding a solution of blue-stone to one of arsenius acid, washing and drying the resultant bulky precipitate. It is used in making blue fire. The kind used for green paint is entirely satisfactory for fireworks making.

COPPER CARBONATE.

This substance is also used for making blue fires but better effects are obtained by the use of other copper compounds, with less trouble. The native carbonate is almost useless for fireworks purposes but the precipitated is easily obtained from dealers in pyrotechnical chemicals or can be made by adding carbonate of ammonia to a solution of blue-stone. Chertiers Copper is made by carefully adding aqua ammonia to a solution of blue-stone, evaporating and chrystalising. Black sulphuret of copper, black oxide of copper and various other copper compounds are occasionally used. The author has obtained the best results with copper ammonium chloride, and calomel is unnecessary with this salt.

COPPER SULPHATE.

For most purposes where a good blue was required for exhibition purposes the older pyrotechnists used this salt but owing to its being a sulphate great care must be used in mixing it with chlorate of potash and a separate sieve should be used for mixtures of these substances, which should not be employed in any other work. It costs usually 10c to 15c lb. Mixtures containing it must not be stored but used promptly after making. Exposure to moist air oxidizes this and releases sulphuric acid. This can be obviated by using potassium perchlorate but the resulting mixtures are much more difficult to ignite in the form of stars etc.

ANTIMONY.

Metalic or Reglus antimony, when finely powdered in an iron mortar is used in making white fire. It may be had from machinery dealers at 6c to 7c lb.

ANTIMONY SULPHURET.
(black)

This may be obtained from drug dealers and costs 6c to 8c lb. but is often so impure as to be entirely useless. If 70% pure it is still servicable for pyrotechnical purposes and is used for making white fire, maroons and smoke effects. Red and orange sulphurets are also sometimes used. The compounds are poisonous.

RED ARSENIC, (REALGAR)
$As\ S_2$
ORPIMENT
$As\ S_3$

These minerals come mostly from Hungary, in iron

kegs of several hundred pounds and range in price from 6c to 9c lb. for the powdered. They are useful in making white stars, especially as these take fire far more easily than those made from antimony. Arsenic compounds are also used for making yellow smoke in day fireworks.

ALUMINUM.

When, about 50 years ago, it was found that a star of unusual briliancy could be produced by the use of magnesium this metal suddenly came into considerable demand in spite of its then cost of $75.00 lb. About the time that its price was reduced to $5.00 lb. it was found that aluminum was in every way better and cost little more than $1.00 lb. in fine powder. It can now be gotten from most paint dealers, in 1 lb. cans or papers at 60c lb. Aluminum powder should be 95% pure. It may contain 2% fatty meterial and 1% silicon. The fine should pass through a 100 mesh sieve and the flake, through a 50 mesh sieve.

CALOMEL.

This is used to deepen the color of fires when they are not sufficiently deep without its use. It has been found that the chlorides of metals give the best spectrum but chlorides are not usually practical for fireworks making so the addition of an easily decomposed chloride to fireworks compositions is to produce a chloride at the moment of combustion, thereby acquiring the desired result. Finely divided metals also take fire spontanously in chlorine gas and the great heat thereby produced probally causes the increased depth of color. Calomel is made in Philadelphia and elsewhere and costs 65c to 70c lb. but this price has recently been advanced.

AMMONIUM CHLORIDE.

Sal Ammoniac.

This is sometimes used as a substitute for calomel

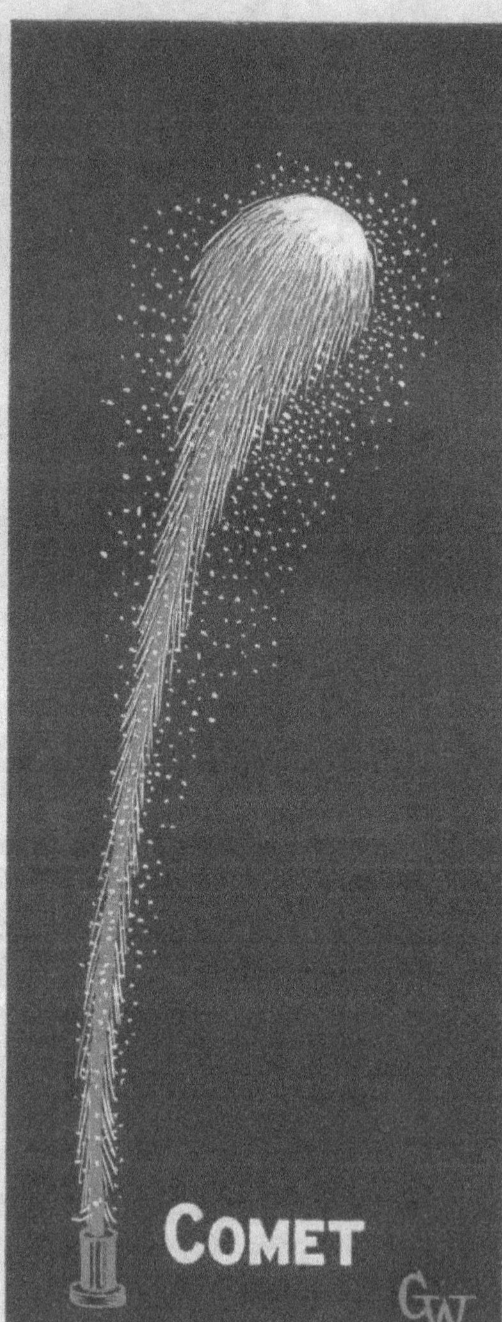
COMET
GW

but its affinity for moisture seriously interferes with its general use. The crystalized salt is almost useless.

DEXTRINE.

In all the old works on pyrotechny, either a solution of shellac in alcohol or gum arabic in water is used to bind compositions for making stars and other similar purposes, but at present, in most cases the necessary amount of dextrine is added at once to the mixture and then nothing but water is needed to form it into the desired objects. Dextrine also improves the color of some fires and it may be advantageously used in place of glue for light work. Potato dextrine usually comes in sacks of about 200# and costs from 2½c to 5c lb. When used for gumming rocket sticks, tabs etc. it is simply mixed with water to the desired consistency. The light brown #152 is most suitable for pyrotechny.

GLUE.

Several forms of glue are used in fireworks making. For attaching lances to frame work a good grade of carriage glue is best. For attaching mine bottoms etc. to the cases cheap carpenters glue will suffice. For placing shell fuses and securing the ends of cannon crackers, good liquid glue is most convienent.

GUM ARABIC.

In powdered form this is used in some star compositions, especially for making Japanese Stars. It is also used in "Son of a Gun" composition.

STEEL FILINGS.

Cast iron borings etc.

A beautiful scintilating effect is produced with steel filings used in various ways. The Japanese make a little tube of twisted paper, at one end of which is a composition which when lighted produces a glowing bead of molten flux. The balance of the tube contains steel filings, which when reached by the fused bead, burst into feather like flashes. In other countries steel filings are added to gerbs, fountains and driving cases with resulting brilliancy. A beautiful waterfall effect is produced by charging from 50 to 200 cases 2 inches in diameter and 12 inches long with a composition containing cast iron borings. These are fastened to scantlings at intervals of about 15 inches, each scantling holding about 16 gerbs. These are matched and hoisted to a wire cable some 50 feet above the ground. When burned the effect is most realistic as the arc of the suspended wire gives just the right curve to cause the appearance of perspective while the roar of the burning gerbs is also characteristic of Niagara Falls as the fire from the iron borings drops to the ground.

The best steel filings for gerbs is known as "needle steel". This resembles broken sewing needles but is really a by-product of some turning or planing operation. The steel filings from saw filing shops are quite good provided they are the result of hand filing and not the particles thrown off by emery wheels, which are useless for pyrotechnical purposes. When steel filings are added to gerb compositions, the saltpeter quickly attacks them, frequently causing the gerb to become quite hot. The steel is rusted and this action practically destroys its usefulness. To prevent this the steel must be coated in some way that the saltpeter cannot attack it which may be accomplished as follows:

In an agateware saucepan place a piece of paraffin and carefully melt it, heating as much as possible without permitting it to smoke. To this add clean steel filings, as much as the paraffin will thoroughly coat. There should be no surplus of paraffin but just enough to completely cover each filing. Shake the pan and stir frequently while cooling to prevent the filings from caking. Steel filings are also used for stars in rockets and shells.

CLAY.

This is used for closing the ends of most cases as well as choking them when they are not crimped. Most any kind of clay will do. It must be thoroughly dried, pulverized and sifted. Before using, it may be slightly dampened. For convienence where a large quantity is required, powdered fire-clay in barrels may be used as this saves the rather tedious job of drying and powdering.

GUN POWDER.

This is used in all grades from Dupont FFF Rifle to the coarse grains as large as cracked corn, for shells. A slow burning powder is preferable for a driving charge as it reduces the liability of shells bursting in the mortar

MEAL POWDER.

This article is used considerably in display work for gerbs etc. and in shells and rockets as a blowing charge. It is generally supplied in 25 lb. wooden kegs but is sometimes difficult to obtain. In that case some pyrotechnists make a fairly good article themselves, as follows:

Mount a 50 gallon wood barrel on two uprights so that it will revolve freely on centers fastened to the heads. On one center attach a crank and cut a hole (closed by a suitable plug) into side of barrel for putting in and removing the necessary ingredients. Place in the barrel 300 to 500 lead balls about one inch in diameter. When it is desired to make meal powder put into the barrel a thoroughly mixed composition as follows:

Saltpeter, double refined	15 lbs.
Willow charcoal	3 "
Sulphur flour	2 "

The barrel is now revolved for about 500 turns. The longer it is turned, the stronger the powder will become Great care must be exercised to see that no foreign matter such as nails, gravel etc. find their way into the barrel as this might result in an explosion.

New Ingredients.

Some years ago powdered magnesium was added to the ingredients used in pyrotechny and very fine bright effects were produced with it. Just when its high price and its affinity for oxygen, causing it to decompose the chemicals with which it was mixed, made pyrotechnists look at it askance aluminum came on the scene. Added to stars and torches it greatly increases their brilliancy and beauty. Exquisite water fall effects are produced with it as well as comets, tailed stars and intensly bright flares. Large quantities of finely divided aluminum (pyro aluminum) are used in the new "flash crackers" and the same composition is used in maroon shells. Besides increasing the report it gives a startlingly bright flash to the explosion. Being unaffected by water it is likewise much safer than magnesium but care should be used in handling it because as, before mentioned all finely divided metals are liable to explosion when in contact with oxygen producing chemicals. Rubbing into it a small amount of vaseline seems to reduce the danger of accident.

Picric acid is another valuable ingredient in fireworks making. When added in small quantities to colors it deepens them and increase their brilliancy without making them burn much faster. Also beautiful colors can be produced with it, almost free from smoke. But it must always be kept in mind that picric acid (tri-nitrophenol) is a first cousin to T N T the tremendous explosive force of which is only too well known. For this reason it cannot be used in shells as stars made with it will detonate when confined, instead of burning. Another effect for which large quantities of picric acid were used until some years ago when a fatal accident occured in a factory employed almost exclusively in making them are the amusing "whistling fireworks". Picrate of potash has the peculiar property of emiting a shrill whistling sound when rammed tightly and burned in a small tube. If made in small quantities and carefully handled it seems to be reasonably safe but the result of a barrel of it accidently detonated can be readily imagined.

Still another substance producing a most beautiful effect when fired from specially prepared rockets, as will be explained later in detail, is phosphorus. It is with this that the so called "liquid fire rockets" are made and a more beautiful display than these does not exist. They consist of an intensly yellow flame melting as it falls through the air, breaking into myriads of incandescent particles with a heavy background of white smoke. Obviously, the greatest care must be exercised in its use as phosphorus burns, even when very small are most painful, but when properly handled it is no more dangerous than many other articles.

Considerable quantities of phosphorus are also used in the manufacture of the article variously known as "spit devil" "son of a gun", "devil on the walk" etc. but deaths of children by poisoning on account of mistaking these tablet looking contrivances for candy and eating them has caused their restriction in some states.

Amorphous phosphorus is the base of most of the toy torpedoes in use. Fulminate of silver was used almost exclusively for this purpose 40 years ago but only a small amount is used now owing to its very sensitive nature. However, its method of use and preparation will be given later as a matter of record.

Zinc powder is used to some extent for making what is known as Electric Spreader Stars. These produce an original effect, breaking up while burning into many small bluish-green particles. These, being propelled with considerable force give the appearance of electrical discharges, hence the name. On account of the explosive nature of zinc dust the making of this star must be done with caution and reserve, until it is well understood.

An effect that is always beautiful, is easily produced, is perfectly safe under all circumstances and is susceptible of an infinite variety of uses is the Japanese or lamp-black star. The well known Willow Tree rockets and shells are made with it and it may be used as garniture for colored rockets, mines etc. An unusual fullness is given to any article to which a small quantity of Jap. star is added.

PART II.

MANIPULATIONS

The handling of explosives, naturally, is never entirely free of danger. No more so is electricity, gasoline and many other things in daily use yet many persons have devoted long lives to the making of fireworks without having an accident. Even with the greatest care, however, accidents will occur to both those employed in making fireworks as well as those burning them. It is here endeavored to point out the most fruitful sources of accident though obviously it is impossible to foresee every instance in which some carelessness or unknown factor may bring on disaster.

First, always keep separate places, a considerable distance apart to be used for making so-called "plain mixings" as rockets, romancandle and gerb composition containing sulphur, and the colored mixings containing chlorate of potash. Separate sieves and utensils of every description must be employed and those working in the "plain" sections of the factory must not go into the rooms of those in the "colored" sections.

Second, keep in mind that very slight friction will sometimes start the burning of mixtures of finely divided chemicals. Star composition has been known to explode while being sifted, by scratching the brass wire sieve bottom with the finger nail, while rockets have taken fire from the brass solid rammer striking the top of the spindle while ramming.

Third, finely divided metals, when in contact with chlorate of potassium sometimes take fire suddenly. While fortunately this is seldom the case it must not be lost sight of. Even steel filings and iron borings frequently become quite warm when mixed with saltpeter etc. and rammed into gerbs. Fire is said to have occured from this action. The prevention of this has been explained under "Steel Filings".

Employes in the mixing and ramming rooms of factories should be required to wear rubber shoes while at work and a constant source of danger is the carrying of matches. This cannot even be controlled by requiring the employes to change their clothes in the factory before going to work and having them wear garments without pockets as they will sometimes slip out for a smoke during rest hours and have matches secreted somewhere about their persons.

Small buildings should be supplied, about 12 feet square and not less than 50 feet apart for all those engaged in mixing and ramming operations as well as for those making stars and as much as possible have one person to a room. Doors should be placed at both ends of work rooms and should always open to the outside with no fastenings on the inner side but held closed, if desired, by spring hinges. Fire buckets, inspected daily should be on each building, supplemented by fire hose convienently placed for emergency.

The most successful method of reducing the liability of serious accidents to a minimum is to keep at all times the least possible amount of composition on hand in the work rooms and to remove to storage or finishing rooms all rammed articles as quickly as they accumulate.

Long experience has shown the following list of mixtures used in pyrotechny to exhibit the characteristics following each and are to be handled accordingly.

SALTPETER, SULPHUR, CHARCOAL and LAMPBLACK seem to be the safest combinations with which we have to deal, accidents with these accur only when a spark has been struck in some manner and brought in contact with mixtures of them.

BARIUM NITRATE with sulphur, saltpeter and aluminum are not reported to have caused any accidents.

BARIUM and STRONTIUM NITRATES, ALUMINUM and similar substances in combination with POTASSIUM PER-CHLORATE have been found to be among the safe mixtures even when mixed with sulphur and gums.

Barium nitrate and Strontium nitrate when in combination with potassium chlorate and shellac or other gums form a sensitive mixture and this condition is largely increased when powdered charcoal is added. Care is therefore urged to avoid all unneccessary friction when handling same.

BARIUM CHLORATE yields its oxygen quite readily so it is to be handled with great care in compounds containing shellac and other hydrocarbons.

ALUMINUM powder in mixtures with potassium chlorate, barium nitrate and shellac or other carbon sources are classed as "hazardous".

Mixtures of potassium chlorate with sulphur, sulphates or sulphides are to be avoided at all times. Chlorates should never be mixed with ammonium salts as this combination is said to be liable to spontaneous combustion. See "Marshall on Explosives".

Potassium bi-chromate and permanganate are to be handled with care, especially in combination with finely divided metals.

Mixtures of picric acid and chlorates are too sensitive for ordinary use. See "Marshall on Explosives".

Compositions of potassium chlorate and phosphorus must never be mixed except under water. Phosphorus alone must never be removed from water for more than a few moments at a time, and then handled so as to avoid all friction.

FULMINATES of MERCURY and SILVER should only be handled by experts.

When care is urged in the foregoing combinations it does not refer to small amounts used by a person of ordinary judgement but more specifically to amounts of 10 to 100 pounds when thoughtlessly handled as so much sand or cement.

When experimenting with new substances use the smallest possible amounts of the component chemicals until the entire safety of the mixture is assured. Before using considerable quantities of new mixtures they should be subjected to exhaustive tests as friction, percussion, detonation and moisture with subsequent drying. Also their flash point should if possible be ascertained with suitable apparatus for this purpose.

MIXING.

The first operation in fireworks making and I may say the most important is mixing. Chemicals are so well made now and can be so easily obtained in a powdered state that long articles on purifying, powdering etc. are unnecessary. All chemicals should, of coarse, be obtained of the best quality procurable at a reasonable price and as finely powdered (as a general thing) as possible but chemically pure drugs are not necessary.

For mixing on a small scale, round brass wire sieves are the best. For lances and the more particular work #22 to #26 mesh may be used while for plain making #16 to #18 mesh is suitable. If 25 lbs. or more of composition is to be mixed ordinary painted wash tubs are most convienent and the sieves should be made so as to just fit inside the upper edge of same while for mixings of from 100 lbs. up troughs are often used. For these, the sieves are made square and fit just inside the troughs, same as with tubs. Mixing machines are sometimes used for bright work or mixings containing no chlorate of potash but they are too dangerous for colors.

With the plain mixings, the coal is weighed first and put into the bottom of the tub; then the sieve put in place and the sulphur, saltpeter etc. pushed through it. When everything is sifted, bare the arms and mix well in every direction. Place the sieve on another empty tub of same size and sift from the first tub into the second one, a scoopful at a time. When all has passed through for the second time repeat once more into the first tub, mixing between siftings and after last sifting. For ordinary compositions this is sufficient but some mixtures are passed four or five times through the sieves.

In colored mixings more care must be observed and each ingredient sifted separately the first time, except the shellac, coal etc. which can be put right into the bottom of the tub. Never throw the chlorate of potash on the sieve at the same time with dextrine or other hydrocarbons but sift the potash first and add other salts one by one. Great care should be taken never to let the fingernails strike the sieve while sifting as it is very easy to "strike fire"from such causes, with disastrous effect as sharp star compositions in a loose state are almost as explosive as meal powder. Special mixings will be described when we come to the compositions requiring them.

CASE ROLLING.

This is the next most important operation of the business and the one requiring probably the most mechanical skill judging from the time required to learn it and the comparatively small number of really good case rollers to be found in most factories.

All kinds of fireworks require a case of some kind except tableau fires. A good case must be tightly rolled and almost as hard as iron. The best arrangement for case rolling is a sort of large desk made of tongue and grooved flooring tightly joined and firmly nailed to sills of about 2 inches thickness and tapering from 2 inches in front to 6 or 7 inches in the back so as to form a gentle rise from front to rear. According to the work to be done the rolling board may be made from two to four feet wide. See Fig. 1. A marble slab also is good for rolling rocket cases.

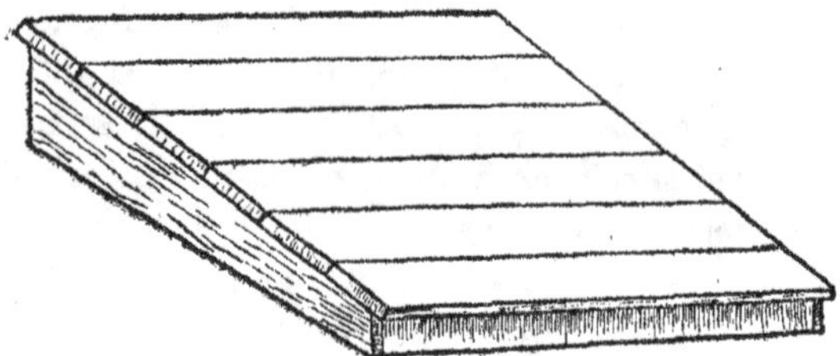

Fig. 1

Most cases are rolled from strawboard, featheredged, The best is made in Elbridge, N. Y. and comes in sheets 26" x 38" with weights varying from 40 to 150 sheets to bundle of 50 lbs. For rockets, two or three turns of hardware or cartridge paper are used first, backed up by five or six turns of strawboard. The cartridge paper being waterproof swells and contracts but little in rolling' while the strawboard, being absorbent swells considerably; therefore when the strawboard is rolled on the outside of the case, it contracts in drying and is shrunk on making a very firm case. Heavy manilla and so called cotton sampling paper also make good rocket cases if carefully rolled but as these shrink considerably in drying, the ramming tools are liable to stick unless specially adapted to this kind of paper. The recently produced "Kraft" paper should make an excellent case though I have never used it. There is also a greyish rag paper which is extensively used for candles.

The best strawboards and pulp boards cost about $25.00 per ton.

The lightest cases used in fireworks making are lance cases. Some pyrotechnists use poster paper of different colors, corresponding to the color of the composition to be rammed into them, while others use linen paper.

Colored paper has the advantage of making lances easily distinguishable in case the boxes containing them become mixed. On the other hand it requires keeping a larger stock of empty cases continually on hand which is sometimes inconvienent. Linen paper is much stronger and only one kind is required, the different colors of lances being kept separated by having boxes for them with the colors marked on the outside.

Lances are made from ¼" to ⅜" in diameter and from 2½" to 4" long. Generally speaking, the greater the diameter, the less need be the length. I generally use a lance ¼" diameter made of ribbed linen paper 17" x 22" about 16 lbs. to the ream, cut in four, the smallest way or across the ribs and six times the long way or with the ribs. This makes 24 cuts from each sheet 3⅝" x 4½". Now procure a brass or copper tube with an outside diameter of ¼" and some good paste. Take a little bunch of say one or two dozen sheets and lay them squarely before you on the rolling board so that the 3⅝" sides are at top and bottom and holding them down tightly with the left hand, rub them gently toward you with the thumb nail of the right hand so that each one will slide about ¼" below and to the left of the one under it. Apply paste to these edges, lay the tube now on the top sheet about ¾" from the bottom of same and ⅛" from left or pasted edge. With the ends of the fingers of the right hand bend the lower edge around the tube, laying over about ¼" and roll to upper pasted end. Then with a turn of the fingers twist the bottom in. The bottoms should not be made too solid and if even a little hole is left in them it will be easier afterwards to stick them on the pins. Sometimes when an exhibition is made on the grounds and not subject to much handling, the lances are made without any bottoms. They may now be thrown lightly in a basket or sieve to dry. These operations while very simple are quite hard to describe and a few moments of practical demonstration will go farther than several pages of description.

Pin wheel cases and match pipes are rolled in a general way the same as lances, except that no bottoms are made to them and brass or steel rods are used instead of tubes. The most convienent size for match pipes is one yard in length and ¼" diameter. Use a good quality of manilla or kraft paper 24" x 36", 20 lbs. to ream. The quire is cut the longest way of the sheet, into strips 4" to 5" wide. A steel rod four feet long is the best for rolling them. Pin wheel pipes are usually made 12" long and 3/16" diameter. Sometimes one end is made slightly funnel shaped by pasting a strip of paper 6" long and 2" wide at one end tapering to ½" at the other, rolled around the end of the rod. Rolling match pipes properly is one of the most difficult operations to master, in the business. It is therefore advisable to begin on shorter pipes until practice is acquired.

ROMAN CANDLE CASES.

These are also somewhat difficult to roll. It is almost essential to have feather edged boards for this work and preferably strawboard. The sheets for one to four ball candles are pasted entirely over with rather thin paste. From six ball up, only about 4" on each end of the sheet should be pasted on both sides. The manner of proceeding is to lay a sheet on the rolling board, pretty well up near the top, and upside down. Then with a 4" flat paint brush apply thin paste quite heavily on about 4" of the top of the sheet and about the same amount on the bottom. Now place another sheet on top of this but about one inch lower down, so that an inch of the first one extends beyond the next on top of it.

Paste as before and repeat the operation until a dozen or more sheets are in the pile. Now reverse the entire lot at once so that the former bottom one will be on top. Paste over the bottom and top edges of pile now exposed and rub off surplus paste with a scrape of the rod and you are ready to begin rolling. (Fig. 2).

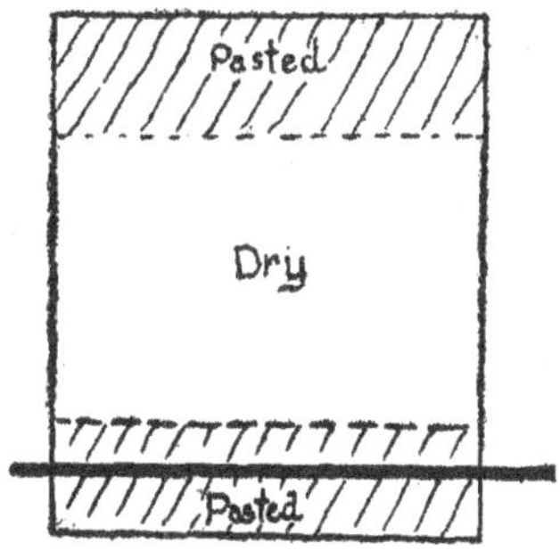

Fig. 2.

Lay the rod across the pile about 3" from the bottom. Lift bottom edge of first sheet, lay it over rod, draw rod with paper around it back, until edge of strip is on top of rod and by sliding the fingers along the rod and edge of sheet until same sticks firmly to it, for its entire length. Now roll firmly along, one hand following the other until the whole sheet is rolled up, care being taken that the case does not run to one side. By a quick backward twist of the rod it may now be removed from the former and placed on rack for drying.

The diameters and lengths of roman candles has been changed and reduced so often of late years that no standard of sizes can be given but the following will be found to be as useful as any for the average work and may be used comparatively. Special sizes may be easily adapted to the required circumstances. When cutting paper for candles and other cases as well care should be taken to always cut so the case rolls with the grain of the paper and with the feather edge at the top of the sheet.

NO.		LENGTH	BORE	SIZE OF SHEET	NO. OF STRAWBOARD
1	ball	4½"	5/16"	4½" x 6"	140
2	"	5½"	"	5½" x 7"	"
3	"	6½"	"	6½" x 8"	"
4	"	8½"	"	8½" x 10"	"
6	"	12"	⅜"	12" x 13"	"
8	"	15"	"	15" x 16"	"
10	"	17"	7/16"	17" x 20"	"
12	"	19"	"	19" x 20"	"
15	"	22"	½"	22" x 26"	120
20	"	26"	"	26" x 26"	"
25	"	32"	"	32" x 26"	100
30	"	36"	"	36" x 26"	"

Cases for rockets, gerbs, fountains, tourbillions, saxons etc. and the small paper guns used for mines, floral shells etc. require considerable skill and strength for rolling, especially the larger sizes. After seeing a great many case rollers at work and employing at different times their various methods, I have come to the conclusion that the following is not only the easiest but makes the best case.

Procure a small hair scrubbing brush of good quality and long stiff hair. Have the paste somewhat stiffer than for candles. Lay the sheet of strawboard on the rolling board, (in the case of rockets, with the sheet of cartridge paper on top of it.) Now, with the scrubbing brush rub some paste evenly over the cartridge paper, (not as much as for candles) and immediately roll up as tightly as possible, except the last two inches or so. Now paste the sheet of strawboard over as you did the cartridge paper and place the partly rolled case on top of it about 2" or 3" from the end nearest you, seeing that the edges of both are even. Raise the end of the strawboard projecting behind the already partly rolled case and bend it around so it will lay between the part of the cartridge paper left unrolled, and continue rolling, pressing meanwhile the case firmly to the rolling board or marble slab until the case is completed. This leaves a case that is already half dry and when completely so, should be firm enough that it cannot be bent in on the ends with the fingers. The advantages of this method of rolling heavy cases is that the paper, especially the strawboard has not the time to become softened and swelled up as when a number of sheets are pasted down at once, and a tighter, cleaner and more easily and quickly dried case results. If too much paste is used, when the case dries the water from the paste evaporates, leaving the case spongy.

The sizes of rockets vary as much as those of candles consequently the following list can only be used approximately:

Size	Length	Bore	Length Sheet Strawboard	No. Stw. Brd.	Hdw. Paper
1 oz.	3½"	⅜"	10"	140	
2 "	4"	"	13"	"	
3 "	4½"	"	17½"	"	
4 "	5"	½"	20"	120	
6 "	6"	9/16"	13"	"	12"
8 "	7"	⅝"	18"	"	"
1 lb.	8"	¾"	20"	"	17"
2 "	9"	⅞"	26"	140	25"
3 "	10"	1"	26"	120	"
4 "	11"	1¼"	"	"	50"*
6 "	13"	1½"	52"*	"	50"*

* These can be convienently used in two lengths.

Mine Cases.

No.	Height	Diameter	No. Strawboard	No. Pieces
1	4"	1½"	140	1
2	4¾"	1 11/16"	120	1
3	5½"	2 1/16"	100	1
4	7"	2¼"	100	2
5	8½"	2⅜"	100	3
6	10"	2⅞"	100	4

Floral Shell Guns.

No.	Height	Diameter	No. Strawboard	No. Pieces
1	9"	2 5/16"	100	3
2	11"	2 11/16"	100	4
3	13"	3 3/16"	100	4
4	15"	3½"	100	5

Gerbs.

Length		Diameter
9"		¾"
11"		1"
13"		1¼"
15"		1½"
15"	Niagara Falls	2"

SHELL CASES.

These, though not being rolled (except the canister shell cases) properly come under this division as they are part of the case rollers' business, to make and are composed of paper and paste.

There are two ways of making them. One, roughly speaking, consists of papering the inside of a hole; the other, papering the outside of a ball. We will take a six inch diameter shell as an example to work on as it is the most popular size and same method is employed for all. First procure a perfectly round ball of wood or some other substance, five inches in diameter. We then cut strips of strawboard and tagboard or heavy paper of most any kind, about ¾" wide and 4½" long and paste them on a board, one on top of another, with so much paste between them that they will not stick together but will become soft and pulpy. I have used a sort of red building paper sold in rolls, which made a better case than any other kind I ever used. The strawboard and other paper should not be pasted in the same pile but two piles made, one pile of each kind.

Now smear the ball or mold well with paste so it will be wet enough to keep the paper from drying and sticking to it before shell case can be finished. Then take strips of paper from either pile first and lay them on the mold, beginning on top and running half way down the side. Lay the second strip so it will lap over the first one about ¼" at the lower end and almost over it on top but ½" lower down. The third strip should start still a half inch further down from the top while the fourth strip again starts at the top. This will prevent the shell case from becoming egg shaped. Continue this until the entire upper half of the ball has been covered with paper. Each strip must be firmly pressed down and all surplus paste squeezed out with the fingers. Now repeat the operation as before but using the other paper. The object is to make it easier to see where one layer has begun and the other ended as each layer is put on. Another way is to cut the strips a foot long and after softening with paste as above, lay them on the mold from the top to the middle, tearing off the strip at the required point and letting second and third pieces start half an inch below the other so as not to get the top too thick as explained above.

After the third layer has been put on one should be laid on crosswise, crossing as much as possible the first layers. This process is continued, pressing each strip as firmly as possible until the case is about ⅜" thick at the edges where it is usually thinest, and not over ⅝" on top. If the work has been properly done the half shell can now be slipped off and allowed to dry. When dry the lower edge should be trimmed off with a sharp knife at a point that will make two halves, when put together show a fair sphere.

The other method is to have a wooden block hollowed out so as to have a hole in it 5½" in diameter and a perfect semi-sphere. Or a mold may be made by taking a ball of this diameter, oiling it well and setting it halfway of its diameter into a box of wet plaster of paris. Now then, proceed as before, except pasting the strips inside the hollow instead of on the outside of the ball. This will make a better looking shell and I believe, a stronger one when properly done. The paper may be cut into strips a foot or more long and torn off as they reach the edge of the hollow. In this way all waste is avoided and the rough edge made more even and regular. The strips should be pressed very firmly as the quality of the shell depends on this. If the pressure against the fingers, in rubbing out the paste, makes them sore, a piece of wood about 3" long and 1½" wide, rounded and slightly curved on the edge, may be used as a sort of squeezer. If the work has been well done, the case should be as firm as wood when dried. To assist in removing the wet shell case from the mold, first place in the bottom of it two strips of cloth at right angles with the ends protruding over the sides, long enough to permit pulling the case out by them when it is completed.

When the halves have been evenly trimmed place two together so as to form a sphere and secure joint with a strip of canvas smeared with glue. Then put on one or two more layers of paper. After again drying, bore hole for fuse through one end or better still bore hole through one half, from inside, with a wood bit, before joining the halves.

In addition to these methods very good shells can be easily and quickly made where hollow balls of zinc, tin or wood can be obtained. The wood half balls need only to be well glued together and they are ready for use. Those of zinc and tin require to be papered just as directed for shell making with round mold except that entire ball is papered until it is about ¼" thick for 6" shell and ½" for 10" shells. Others in proportion.

The cases for cylindrical or canister shells need no detailed description as they are made just like any other heavy case. A former of the required size is procured and the case rolled thereon just as for a mine, of straw-board or other paper. When thoroughly dry the wooden heads or plugs are fitted, nailed in with 1" iron brads, or well glued if made with a flange and carefully sealed all around with several thicknesses of good manilla paper.

DRYING CASES.

For all cases more than 6" long, racks are most convienent for drying them. These are made of strips of ⅞" x 2" cypress or other light wood, suitable for supporting them. The longer the cases, the farther apart the strips should be. When filled with cases they should be moved to a well ventilated room or covered platform. If placed in the sun they will be badly warped when dry. (Fig. 3) Center and end strips are ⅞" x 3".

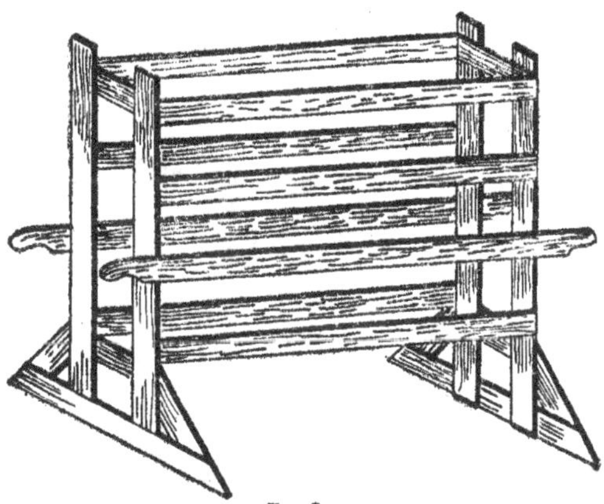

Fig. 3.

Small cases may be thrown into sieves 2 feet wide by 4 feet long, made of 1" material, 4" deep and the bottoms covered with galvanized hardware cloth of ½" mesh.

When cases are stored away care should be taken to protect them from roaches and mice as these are attracted by the paste.

FORMERS.

All paper cases are rolled on formers of one kind or another. For rockets, gerbs etc. these may consist of hard wood sticks but better formers are made of light brass tubing with an outside diameter equal to the inside diameter of the case desired. They should be one to two inches longer than the intended case and fitted with wooden handles to enable them to be easily removed when case is rolled. (Fig. 4)

Fig. 4

Rocket case Former

Mines etc. are rolled on wooden formers, the ends of which are turned down to convienient size to fit the hand. Roman candles are rolled on rods of machine steel while match pipes and pin wheels are rolled on thin brass or steel rods. Lances, on small brass tubing.

PASTE.

Without this simple article, I doubt if any great amount of modern fireworks could be produced, as it is in almost constant demand in every department of the factory.

Ready made paste is now so easily and cheaply obtained that few persons care to bother about making it but for those not so fortunately situated the following is a standard method:

> Mix 4 ozs. wheat flour with 8 ozs. water and ⅛ oz. powd. alum, rubbing until free of lumps. Pour this slowly and with constant stirring into:

16 ozs. boiling water to which has been added,
5 drops carbolic acid,
5 drops oil of cloves,
2 grains corrosive sublimate.

When cold it should be ready for use.

Another method which I am about to describe, I think, not only the best and simplest but requires nothing for preserving and if properly made, will keep for a month in winter. The process consists mainly in allowing the batter to sour before cooking; and cooking by adding boiling water instead of placing directly on the fire where it is likely to get lumpy or overcooked. The following details are for making paste in lots of three or four buckets per day.

Procure two deep wooden tubs of about 20 gallons capacity. Buy a barrel of the cheapest grade of flour you can get. Samples or sour and wormy flour will do if it is not adulterated. Put 2 or 3 bucketfuls of flour into one of the tubs and add water, stirring meanwhile with a paddle until well mixed and about as thick as is convienent to handle. It does not matter if it is lumpy as these all come out in the souring. When the tub has been filled not more than one third full allow it to rest in a warm place (about 90°F) for two or three days by which time fermentation will have set in. When the fermentation is complete the flour will have settled as a heavy batter in the bottom of the tub with a sour brownish liquid over it. Pour this off and fill several buckets about one third full of this batter. Now have a water boiler of about ten gallons capacity with faucet in bottom, on a gasoline stove or furnace and when this is filled with boiling water place one of the buckets of batter under the faucet. Open it and while the water is running in stir it briskly. The contents of the bucket will at first become as thin as milk but as it begins to fill it will gradually thicken until it can hardly be stirred if all the details have been correctly followed, and a bucket of clear, clean and very sticky paste, free from all lumps will be the result. The other tub may be used alternately with the first for souring batter while that in the first is being used for paste making. This paste, having been soured before cooking cannot sour again and will not become watery.

Glue and dextrine are sometimes added to make paste bind better and alum, bluestone, salicylic acid etc. to preserve it but these are all unnecessary if made as above.

CRIMPING.

Sometimes gerbs, etc. are choked or crimped to reduce the opening, in place of using clay. This is done

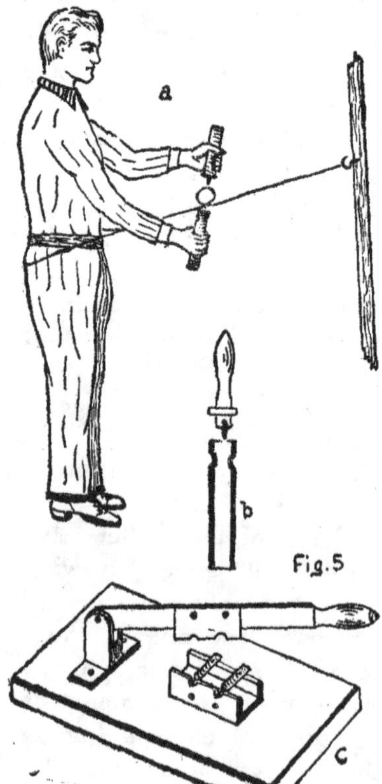

by taking a turn of strong string or piano wire around the case while still wet, about ¾" from the end and drawing tightly while turning the case slightly so as to make a neat job. One end of the string should be tied to a wall or some unyielding object while the other is passed around the body (a) (Fig. 5). A nipple with a short point slightly smaller than the desired opening to be left in the end of the case, should be inserted about ½" before drawing the string so that the end of the case will be kept open and crimping neatly done (b). A mechanical device made by a Cincinnati machine works does the work very neatly and much quicker than the string process, (c) (Fig. 5).

RAMMING.

As this operation will be described in detail under each of the articles to be rammed as we come to them, only a few general directions will be given. All ramming should be done in small sheds as far removed from the balance of the factory as practicable and with one side open toward which the operator should always have

Fig. 6

his back while at work. A stout wood block, either resting on the ground or over a foundation should be used for ramming on. For heavy ramming the best mallets are those made of raw-hide. These are round and range in weight from ½ lb. to 10 lbs. About 2 lbs. is a good weight for the average work. (Fig. 6).

Ramming tools should be made of brass or gun-metal also the nipples, while the spindles for rockets must be made of steel.

Scoops for taking up the desired amount of composition at a time, can be made of tin or any light metal and should be provided in different sizes from about ½" in diameter and 1" long to 1" in diameter and 3" long, with about six intermediate sizes as some compositions work better when rammed in small quantities than others with the same caliber of cases.

RAMMING with ROD and FUNNEL

For all small work such as serpents, saucissons etc. make a funnel about 4" high, 3 inch diameter on top and 5/16" at the bottom, without a spout, (Fig. 7) (a).

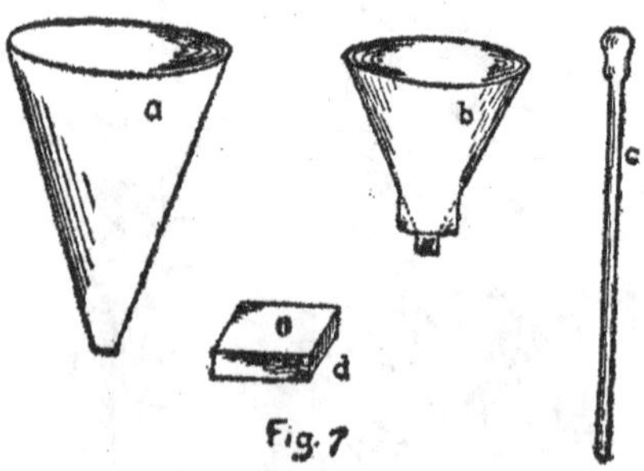

Fig. 7

Procure a rod ¼" diameter and 12" to 18" long according to work to be done. A wooden knob may be fastened to top of same for convienence in ramming (c). In use, a case is slipped on a nipple (d). The funnel, half full of composition, has its small end inserted in top of case and with the rod moved up and down, striking the bottom firmly each time, the composition becomes rammed with sufficient solidity. When case has been rammed to within ½" of top, funnel is removed and a charge of clay is added to stop end, by striking clay a few blows with a light mallet and suitable drift or rammer. The arrangement for lances is somewhat lighter. The funnel (b) is very efficient. It is 2½" diameter on top and 2½" high with a ¾" shoulder on bottom and a spout ¼" outside diameter projecting from bottom for ¼". This, when removed from lance leaves just the proper amount of case empty for priming.

MATCHING.

This is the term used to designate that function of pyrotechny which consists of bringing fire to the various parts of devices as they are burned. In most of the individual articles a short piece of match is twisted in the nosing of the wrapper or fastened otherwise. In set pieces this operation takes on an importance second to none in the art.

Starting at a leader at which a set piece is lighted and which must be long enough to reach from the device when erected, to a convienent distance from the ground so the operator can reach it, the match must lead to every part of the piece.

The matching of lance work is fully described under that heading. In the case of set pieces consisting of gerbs, wheels etc. the gerb is first primed by smearing a little priming on inside of choked end of case. A nosing is put on, consisting of 2 or 3 turns of stout paper rolled around end of case so as to project 2 inches beyond end. About half an inch of the piping is removed from a length of quickmatch. This is bent back, inserted into the nosing and secured by tying tightly with two half hitches. The match is now brought over to the next gerb and bent at right angles over it. At 2 inches from this point it is bent back again onto itself and at point of first bend, again at right angles so as to lead to the next gerb. (Fig. 8). At the bottom of this bend the piping is cut off, bearing match, with a sharp knife, and this portion pushed into nosing of second gerb and secured by tying as before. See (Fig. 8). Candles, wheel cases etc. are treated in the same manner.

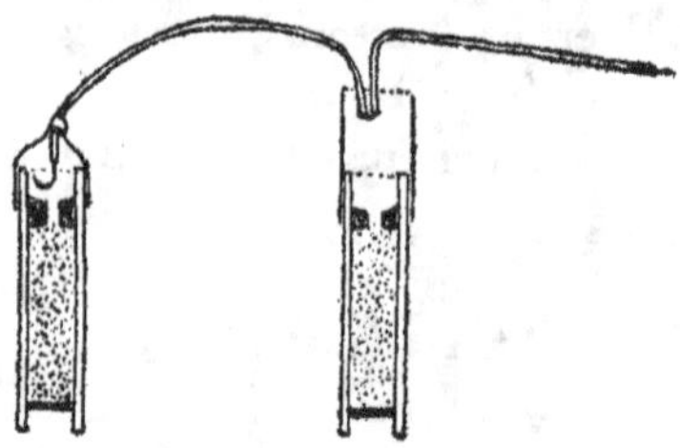

Fig. 8

If a gerb has been properly primed it is not necessary for the match to enter choke as fire will reach it from priming.

It is a good plan to have the leader from which a piece of fireworks is ignited, to run to each section of same, irrespective of the fact that said sections are already connected to one another in the process of matching it, as sometimes a length of match will go out in the center of the pipe, owing to some defect not observed in making it. It is therefore advisable to have the match joined wherever it crosses, as for instance, on top of a lance guarding as much as possible against all chance for one or the other section to fail.

If it is desired to have one part of a piece to burn after the other has been burning a-while, as when candles or gerbs are used in connection with lance work, these gerbs etc. are matched to a separate leader which may be fired by hand after the lances are half consumed, or they may be connected to several lances about half way down so that when they have burned to this point the balance will be lighted automatically. This is done because, lances burning so much longer than candles or gerbs, if all are fired at once, the gerbs etc. would be finished before the lances were well under way, while it is best for the finest effect to come at the end.

PRIMING.

In order to insure lighting, especially in exhibition work, all gerbs, wheel cases, lances etc. are primed which consists of smearing a little moist gunpowder about the mouth of the case. Priming is made by adding water to grain powder in a suitable receptacle until the powder becomes pasty. A little alcohol and dextrine can be advantageously added.

FINISHING.

In factories, where stock or shelf goods are made this is quite an important department. All kinds of fireworks are covered with some kind of colored paper and often stripes and borders are added. Candles, rockets, serpents, small mines and triangles are covered with different colored poster. Flower Pots are usually covered with calico paper while fancy rockets, large mines saucissons, floral shells and fountains are covered with glazed paper, stripes being added where desired. The size of cuts as given here are for use with candles and rockets of the size shown in this work. They are usually 2" longer than the article to be papered if it has to be matched at one end and tucked in at the other; 1" longer where matching only is done and the same length where only the case is covered as in mines etc.

1 oz.	Sky	Rockets			3" x 5"
2 "	"	"			3" x 6"
3 "	"	"			4" x 6½"
4 "	"	"	with	cone	4" x 6½"
6 "	"	"	"	"	5" x 7"
8 "	"	"	"	"	6" x 8½"
1 lb.	"	"	"	"	6" x 9½"
2 "	"	"	"	"	7" x 10½"

1	ball	Roman	Candles	2" x 6"
2	"	"	"	2" x 7"
3	"	"	"	3" x 8"
4	"	"	"	3" x 10"
6	"	"	"	4" x 14½"
8	"	"	"	4" x 17"
10	"	"	"	5" x 19"
12	"	"	"	5" x 21"
15	"	"	"	6" x 24"
20	"	"	"	6" x 28"
25	"	"	"	6" x 34"
30	"	"	"	6" x 38"

WRAPPING.

To make a good neat, tight and strong bundle is about as important and I may say difficult to learn as any other part of the fireworks business.

Roman Candles from 1 to 4 ball are packed 3 dozen in a bundle. From 6 ball up, 1 dozen in a package. The packages of 1 dozen are made in two forms, viz: four sided and six sided, (Fig: 9) (b) and (a) respectively.

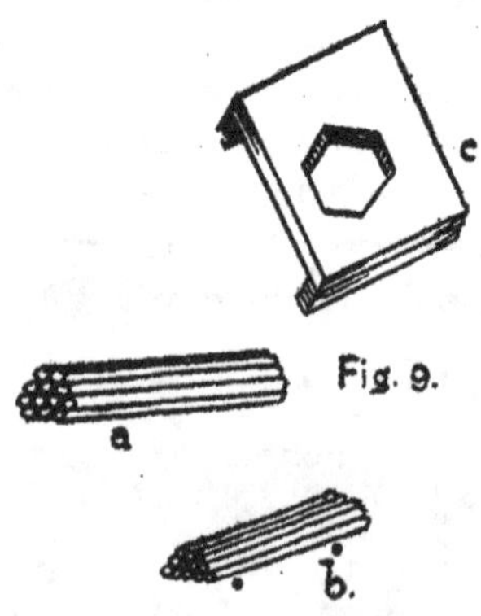

Fig. 9.

To make the four sided package of 1 dozen 8 ball candles lay 5 on the bench in front of you so the candles run paralel with the bench. Mark the space they cover and fit into top of bench four wood pins about one inch of which project above, two on each long side of the space occupied by the 5 candles, so same may be easily laid between them. See dots, (Fig. 9), (b). Now cut a sheet of manilla paper 19" long and 14" wide and lay this between the pins just as the candles laid before, and replace the five candles, now on top of the paper between the pins. On top of these five place four more and on top of them three. This makes twelve. Draw the paper tightly over them and fold it like a druggist makes a bundle. Now close the ends as follows: with two fingers press the top of the folded paper over the end of three top candles; then, holding same down with both thumbs, fold in the two sides of package with the first and second fingers of each hand at the same time; then holding these folds with the left hand, lift the opposite end of the bundle with the right which will cause the bottom to fold itself over the other folds. Now, with a brush dipped in thick paste give the end a daub on the last fold and while the bundle is still standing on this end fold the top end the same way. Before folding the last fold give it a daub of paste as you did the other end. Lean against the wall and place a paper weight or tile on top of end to hold it in place until dry. After a quantity are packed like this and dried the labels are affixed.

To make the six sided bundle a person must first learn to form the candles in the hand. Count out one dozen candles and encircle the bunch at one end with both hands. Now work them about (this is hard to describe on paper) until they form a triangular bunch (a) with three candles on each side of the triangle. When this has been accomplished, lay them on the wrapping sheet (cut as described above, though preferably wider) holding them lightly so they retain the triangular form. To get the paper around them without having them to fall in a heap is still harder to describe and equally hard to master though easy enough when learned. Once the bundle has been gotten to the wrapping sheet one hand is sufficient to maintain its form so with the other lift the side of the sheet nearest you and bend it partly around the package so you can hold it while the other hand is released long enough to enable it to take the paper on that side. Straighten and flatten it well on over the candles and begin rolling up the bundle until the other edge of the paper is reached. Paste this edge and lap it on the bundle and you are ready for the corners. If the bundle has been properly made, when it lays on one of the faces of the triangle, the top row must be composed of 2 candles; the second row, 3 the third row 4 and the bottom row 3. Now bend the paper down from the top first, then bend in the two upper sides, then the two lower sides and finally, by lifting the bundle, from opposite end, the bottom folds over all the others. A little paste secures it as described above.

The bundles of smaller candles are formed in a wooden former, hollowed out to the size of a bundle of 3 dozen (c) and when it is packed with the required number they are secured with a string perparatory to wrapping.

Short stick sky rockets are nearly always packed in paper boxes. Long sticks are packed as follows: Cut some pieces of #18 iron wire 6 inches long. Then take half or a dozen rockets with the heads all even and work them in the hands until they form as square a bundle as possible and bend one of the pieces of wire around the sticks just below the matches. This should be done with one hand while the other holds the bundle in shape. Now pass another wire around the sticks about a foot from the bottom. Cut some pieces of strawboard as wide as the bundle of rockets on the wide sides and long enough to go completely over the heads and down the other side nearly to the matches. Cut some wrapping paper six inches wider than the rocket head bunch and long enough to go twice around it. Paste the far edge for about one inch and lay the bundle of rockets with the strawboard around it, on the sheet and wrap it up as tightly as possible. Fold in the upper end; secure with a little paste and set aside, heads downward, to dry. Later, the other end may be gathered in and secured with another piece of wire.

Wheels, tourbillions etc. are made into most any kind of a package desired, while mines, fountains etc. are given one or two turns of paper over the finishing to keep them clean.

Serpents, flower pots and torches are packed like roman candles. Blue lights, the same. Fancy rockets are packed heads and sticks separately, the heads in boxes and the sticks loose.

WIRING.

For most purposes annealed iron wire from 18 guage to 20 guage is the most servicable. The easiest and quickest way to use it for wiring rockets, triangles etc. is to cut it in lengths of from 4" to 6" according to the size of the work to be done. A large quantity can be cut at once by using a bench shear and cutting several hundred at one time. Rockets can be quite securely fastened with one wire if a GUM BOARD (Fig. 10) is used. Else two wires are necessary. A gum board is made by taking a piece of ½" board 6" long and nailing pieces of rocket stick around it on three sides on top and one side on the bottom. Put into this about 1 oz. of dextrine mixed with water to the consistency of jelly and it is ready to commence wiring.

Fig. 10

Put a pile of rockets and wire to your left and a bundle of sticks and the gum board to your right. Rub one side of the end of a stick against the bottom of the gum board so a little gum will adhere to it. Lay it with the gummed side against the rocket about three quarters of the way to the cone. Hold it in this way in the left hand and with the right, bend a wire around it about the middle giving one turn on the side of stick. Now, with a pair of nippers give about three more turns cutting the wire with the last turn. If no gum is used two wires are necessary.

TYING.

In doing exhibition work string plays a very important part and the best and most convienent knot for all purposes is the sailors two half hitches. (Fig. 11).

Fig. 11

This is somewhat difficult for most persons to learn. The best way is to practice on a stick. Pass string under stick bringing free end over left of loop; bring it over same again passing end again to left of second loop but between second and first. An ordinary tie of the free ends now secures it permanently. This knot will be found invaluable in matching.

LABELING.

This very easy operation may be still further simplified if done in the right way. Take a board about a foot square. Smear it well on top with thin paste and lay a label on it, face down. Cover this well with paste and place another label on top of it repeating the pasting and putting down of labels until several dozen are on the board. This will soften them so that when taken up and pressed with the fingers or the paste brush against the bundle to be labeled they will adhere firmly and lay flat.

DESIGNING.

When it is desired to produce in fireworks a portrait, a picture of a building, monument etc. or a line of lettering this is first drawn on the floor with a piece of chalk fixed into the end of a stick so that the designer may walk about sketching his picture from the miniature plan as he goes along. The floor is first laid off with a chalk line into squares one foot each way and in multiples of 50 square feet five feet wide and 10 feet long. For instance, if a picture 10 feet high and 20 feet long is desired it is composed of 4 sections 5' x 10' or two high and two wide.

The sketch is now taken and marked off with rule and dividers into 200 equal squares, 10 high and 20 wide corresponding to the full sized squares on the floor. These are numbered along the edge of sketch on top and on one side. The squares on the floor are numbered in the same way. With the chalk now draw into each square on the floor, the same lines as appear in corresponding square of sketch. When this is done, an exact reproduction of small picture will be ready to be placed on frames.

For lettering or lines of wording this is not necessary as design can usually be drawn directly onto the floor with free hand, of the desired size and without enlargement.

PART III

ARTICLES OF MANUFACTURE.

MATCH.

(Quick Match)

This is used for conveying fire to the cumbustible portion of pyrotechnical devices and is distinguished from FUSE by the fact that its effect is almost instantaneous while fuse burns at a comparatively very slow and exact rate. It consists of cotton wicking impregnated with gunpowder and covered with a loose paper piping. As almost every piece of fireworks requires match for lighing it and lance work and exhibition pieces in general are absolutely dependent on good match for their successful operation it is essential to make this very necessary article as nearly perfect as possible. There are several ways of making match which will be classified as the "French System" and the "English System" and candle and rocket match.

French System

Secure two pieces of 1" x 3" lumber and into one edge of each drive a number of 8^d nails for half their length, about one inch apart. Set these pieces up horizontally, with the nail edges uppermost, about 3 feet above the ground, one at each end of a dry shed some 30 feet long. Wicking or cotton cord of the proper thickness can be secured already in balls, as desired. Now get some cotton cord or wicking not less than 24 mesh and fasten the end to the nail nearest the wall, on one of the above pieces of 1" x 3". With the ball of cord walk to the other end of the shed and, drawing it rather tightly, fasten it to the corresponding nail in the other strip of 1" x 3" by taking a few turns around the nail. Leave the ball temporarily at this end of shed, in a light box to keep it clean.

In an agate pan mix 3 lbs. rifle powder thoroughly with 4 ozs. dextrine and add water, stirring with the fingers until all the grains are wet. Allow to stand a few minutes until a small lump pressed between the fingers feels perfectly smooth and contains no more grains. Stir in some more water and a little alcohol until the mixture is about the consistency of mush.

Holding the pan in the left hand, under the first length of cord take up a handful of the powder mixture and work it well into the cotton while holding the pan so as to catch the drippings, and walk backwards to other end of shed. When this is reached go back to the begining of the strand, take some of the powder in the right hand, pass the cord over the first joint of the first finger, place thumb on top of it and again walk backwards toward other end of shed but without working any more powder into the cord, simply allowing it to run through the finger and thumb for the purpose of rubbing off rough uneven places and leaving a smooth well finished surface.

Now take up the ball of cord again and, passing it around the third or fourth nail to the right, stretch a second length to the point of begining, fastening it here also several nails away from the first strand. This is for the purpose of not touching the finished strand while working the powder into the second one. Proceed as with first length and when finished move it to second nail, stretching it tightly into place. Repeat with the following strand until all the powder is used up.

If the weather is dry match will be ready for piping in a day or two. In dry climates gum arabic makes a better match than dextrine but where there is much moisture in the air dextrine is safer. When the match is dry and stiff it may be cut down and the pipes threaded on. Match pipes are made of 20 lb. manilla or kraft paper 24" x 36", cut into strips 4" wide and 36" long, rolled on a ¼" steel rod, only the edge of the sheet being pasted for about one inch. When piping match crease or gather the end of first pipe when in place so next pipe may be slipped over it for about 1 inch.

A simple and clean method of making perfect exhibition match, and one, which to the best of the authors knowledge is original, is as follows:

Make a cup of brass, about 3" diameter, at top, 2½" at bottom and 2¼" high as shown in accompanying sketch.

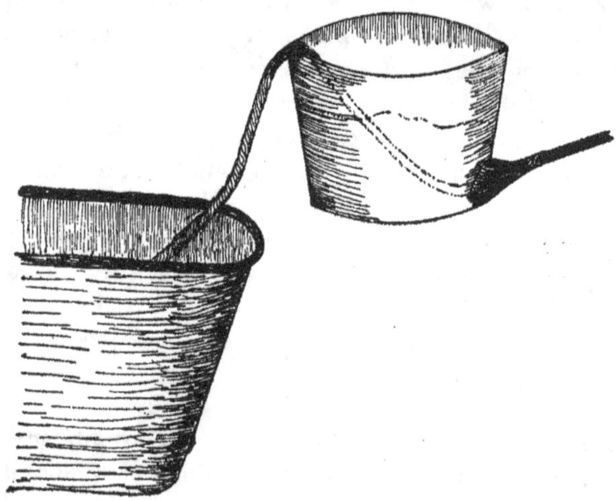

At its bottom attach a spout pointing upward and terminating in an opening ⅛" diameter. A small dish pan, a 2 quart pudding pan and a match frame 4' x 6' complete the requirements of apparatus. Then prepare the following mixtures;

No. 1		No. 2	
Dextrine	3 ozs.	Dextrine	1½ ozs.
Gunpowder	2 lbs.	Gunpowder	1 lb.
Water	2 pts.	Water	12 ozs.
Alcohol	4 ozs.		

Into the dish pan place mixture No. 1, mixing the gunpowder thoroughly with the dextrine before adding water. When powder has completely melted add alcohol, and stir well. Unwind into this about 2 lbs. of good cotton twine of not less than 24 mesh, and with a stick press it well into the powder mixture. Into the pudding pan place mixture No. 2 proceeding as for No. 1. This however, should be thicker; (about like soft putty).

Now take the end of the cord and pass it through the spout of brass can, from the inside; fill the cup with mixture No. 2 and pull through spout enough cord so end may be attached to match frame. Hold the cup in the left hand and revolve the frame with the right, placing dishpan so cord will feed out, over notch in back of top rim of cup and through powder mixture into frame separating each strand by about ¾". If spout of cup fits cord snugly a perfectly round, smooth match will result and if surgeons rubber gloves are worn the hands will not be soiled. Be sure to keep the small cup always full.

English System

Make a light frame of wood like the frame of a looking glass (Fig. 12), six feet long and four feet wide

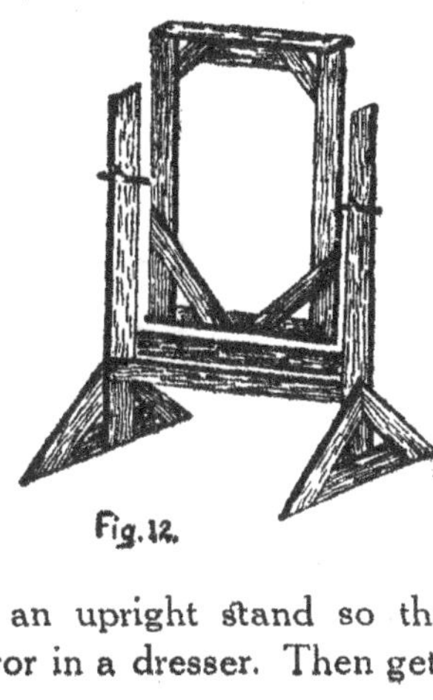

Fig. 12.

and hang it in an upright stand so that it can revolve just like the mirror in a dresser. Then get a quantity of cotton wicking, 24 mesh and unwind it into a thin pan about a foot in diameter and six inches deep. In another similar pan put 2½ lbs. rifle powder mixed with 2½ ozs. dextrine and cover it with 2½ pints water, stirring occasionally until powder is melted; then add 2½ ozs. alcohol and mix well. Pour this over the lamp wicking in the first pan taking care to leave the end of wick hanging over edge of pan so it can be easily found. Begining with this end now run all the wicking into the empty pan, taking care that every part of it is well soaked with the wet powder, a little of which should remain after the cotton is passed for the first time and this may now be poured over the pile of wet wick, pressing or kneading same so as to thoroughly soak every part, when it can be returned to the first pan as before. It is now ready for the frame. Tie the end of wick to one side of end of frame and while someone turns it slowly feed match into it with the strands about ¾" apart. When all of it is on the frame remove from the stand to a part of the floor covered by large sheets of paper and support it over these on four blocks about 3" high, one at each corner of frame. Now take a small sieve of meal powder and dust it carefully over so as to cover it evenly with a layer of powder. The frame of match may now be placed in the sun or elsewhere to dry. Match made by this process is all of one length, viz: 2 yards, and is very round in appearance. It burns fiercely but will not stand as severe usage as that previously described. It also takes longer to make.

Match piping serves the double purpose of protecting the match from injury and making it burn infinitely faster. A piece 20 feet long will flash from one end to the other in less than a second.

ROCKET and CANDLE MATCH.

Match made by either of the two previous methods is too expensive for use with the cheap grade of stock fireworks on the market so a simpler method has been devised for this purpose. It is essentially like the last described process.

Into a small tub put a gallon of starch, well boiled, and stir into it about 15 lbs. of a thoroughly mixed composition made of

Saltpeter	16
Fine charcoal	5
Sulphur	2½

Soak in this, cotton wick of about 5 strands until nearly all the composition is absorbed but about one half inch which should still cover the cotton in the tub. Work it in well and run it on a frame as directed in preceeding description but the frame may be smaller for convienence of handling by one person, as long lengths are not required. Neither does it need to be dusted with meal powder. If well made, however, it will burn freely and serve its purpose completely. When dry it is tied in bunches, 1 to 2 inches in diameter and cut into the desired lengths with a tobacco cutter or large sharp knife.

FUSE.

(Blasting)

This is used in fireworks making, in the production of cannon crackers and to a lesser degree in small bombshells. It consists essentially of a cotton pipe containing meal powder and burns at a rate of approximately 1 inch in 3 seconds. It can be had in sizes varying from ⅛" to ¼" in diameter and from the cheapest painted cotton kind to one heavily coated with gutta percha for under-water work.

It is made by a very ingenious machine which weaves a cotton fabric around a small tube. As this tube is withdrawn its place is taken by the meal powder which is forced in through the opening in the tube. The largest factories are in Simmsboro, Conn.

TABLEAU FIRE.

This is about the simplest form of fireworks at present in use. It is made by mixing thoroughly the necessary ingredients to produce the desired color and heaping it on an iron plate or board, in a pile, so it may be easily lighted. Or it may be put up in tin cans for the trade. Good tableau fire should burn brightly without sputtering and smoke as little as possible. It should take fire easily but never be liable to spontaneous combustion. Lithographed cans may be used as containers, designating by their color the color of the fire they contain and with firing directions printed on them. A small piece of match placed in each can facilitates lighting it.

WHITE FIRE.

Saltpeter	3	12	8	7
Sulphur	1	2	2	2
Metalic antimony	1			
Antimony sulphide	1	1		
Realgar			1	1½

BLUE FIRE.

Potassium perchlorate				24
Potassium chlorate	6	16	8	
Paris green	4		6	
Shellac			½	
Stearine	1		1	2
Barium nitrate	4		7	
Calomel		12	1	
Sal ammoniac	1			
Copper ammon. chloride		4		6
Asphaltum				1
Lactose		6		

In the above color it should be bourn in mind that paris green is very poisonous and a handkerchief should be tied over the nose if it has to be handled much.

RED FIRE.

Strontium nitrate	80	10	16	14
Potassium chlorate	20	4	8	4
Shellac			3	
Red or Kauri gum	12	3		
Asphaltum				3
Charcoal		1		
Dextrine	1			
Fine saw dust	12			
Rosin		1		
Lampblack	1			

If it is desired to make tableau fires more bulky one or two parts of fine sawdust may be added to any of the above recipies without materially affecting the color. If the sawdust will not pass freely through the sieve it may be added after the other ingredients are sifted and mixed and rubbed in with the hands.

PINK FIRE.

A fire somewhat cheaper than the above but inferior in color may be made as follows:

Nitrate of Strontia	48	16	18
Saltpeter	12	4	7
Sulphur	5	2	2
Charcoal	4	1	½
Red gum		3	2
Dextrine			½

This should not cost over 7 cents per pound while the other formulas cost about 9½ cents per pound. A pink light may also be made by substituting lime or chalk for strontia but more chlorate of potash is required and the smoke is greater.

YELLOW FIRE.

Nitrate of baryta	36
Oxalate of soda	6
Sulphur	3
Red gum	5

GREEN FIRE.

Nitrate of baryta	8	9	4
Chlorate of potash	4	3	2
Shellac		1	1½
Red gum	2		
Dextrine		1/16	
Fine saw dust		½	
Sal ammoniac	1		

I have never found anything better or as good as this, therefore give only the one recipe.

SMOKELESS TABLEAU FIRE.

For theatrical or indoor use colored fires are very objectionable on account of the choking smoke they give off. The following mixings give a fire producing very little smoke which quickly dissipates after fire is burned.

RED.

Nitrate of strontia	8
Picric acid	5
Charcoal	2
Shellac	1

GREEN.

Nitrate of baryta	4
Picric acid	2
Charcoal	1

Disolve picric acid in boiling water; add strontia or baryta; stir until cold and dry on filter or piece of cloth.

It should be observed that in all mixings, the formulas cannot be considered absolute as the purity and general characteristics of chemicals differ so much that all mixings must be tested and regulated to the existing conditions of materials, climate etc. If tableau fire burns too slowly more potass; or coal should be added; if too fast, more strontia, baryta etc. In rocket, candle or gerb compositions, saltpeter or meal powder will increase the combustion while coal and sulphur will retard it.

TORCHES.

These may be clasified according to the purpose for which they are intended. Military torches have but one requirement which is that they produce the maximum illumination of the deepest hue of color desired. As these are fully described in special works issued by the government and really form no part of commercial pyrotechnics it will be unnecessary to devote further space to them here. Railway torches or fusees, on the other hand, are the cheapest form of pink light, as anything capable of attracting the attention of the engineer is all that is required. They are usually ¾" diameter and 8" to 12" long exclusive of the handle and burn from 5 to 20 minutes. The following compositions are adequate:

RAILWAY FUSEES

Potassium perchlorate				2½
Strontium nitrate	48	18	16	18
Saltpeter	12	7	4	
Sulphur	5	2	5	2¼
Fine charcoal	4	½	1	
Red gum	10	2		
Dextrine		½		
Sawdust				1

moisten with Kerosene before ramming

Fusees are provided with a slip cap which is used for igniting them. The end of the torch is capped with paper onto which is painted a mixture of

Potass; chlorate	6
Antimony sulfid	2
Glue	1

while the end of the cap is similarly painted with a paste of

Black oxid of manganese	8
Amorphous phosphorus	10
Glue	3

When the cap is pulled off and struck against the end of the fusee it takes fire like a safty match. With some compositions it is necessary to have a little starting fire at top of torch just under the capping or priming under the cap which will suffice to cause easy ignition. (Fig. 13)

Fig. 13.

PARADE TORCHES

Parade torches for campaign purposes, where a cheap grade of fire suffices and where competition urges the manufacturers to produce the largest article at the smallest price, one of the methods is to add 50% of fine sawdust to the mixing. This does not greatly affect the burning of the torch and makes it look twice as large at practically no extra cost. The following is a good formula:

RED.

Nitrate strontia,	30
Potass; chlorate	8
Red gum	7

Saw-dust may be added ad libitum. The torches are usually ¾" diameter and 12" long and should burn, with the above mixing, 8 to 10 minutes.

RAMMING CAMPAIGN TORCHES.

A very cheap method of ramming these torches is to moisten the composition with dilute dextrine solution until it is damp enough to hold together when a handful is tightly squeezed. A dozen torch cases are tied in a bundle and pressed into a pile of damp composition on a slab. It is then moved to a clear part of the slab and jolted firmly against it by lifting the bundle a few inches and jarring it downwards. More composition is shaken in from the top when the jarring is repeated and this continued until torches are full when they are set aside to dry. By this manner a dozen torches may be rammed in one minute. The handles may be attached by a strip of gummed paper 2 inches wide, half of which encircles the torch and the other, the end of the torch handle. The other end of torch is nosed and matched in the regular way, (Fig. 14).

Fig.14 Campaign torch.

A better way of ramming torches is as follows:

Tie the cases in bundles of 12. Place on ramming block and insert spout of funnel (Fig. 15), into one of

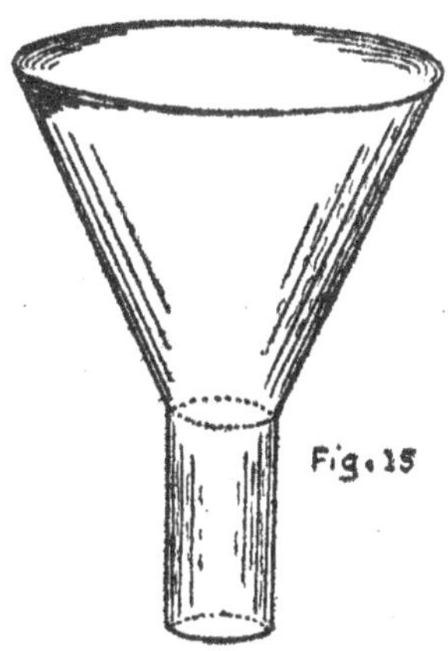

Fig. 15

them. Then pass a suitable rod through it until it rest on block also. Now, with a scoop, fill the funnel with composition and, steadying same with the left hand, grasp the rod firmly with the right and raising it about six inches, drive it with a firm stroke up and down as the composition runs into the case. Continue this operation until case is filled. When funnel is removed the space occupied by the spout will serve for inserting the handle which is done by applying to it a little gum or glue.

TORCHES for CARNIVAL PARADES.

These present the most exacting requirements and the following formulas are the result of more than 30 years of experimenting during which some exceptional mixings have been developed as well as some most beautiful colors which, in modified form, may be used for box stars in shells etc.

Carnival Parade Torches must be of deep color, give maximum illumination, burn slowly and clean, not be prohibitive in cost and give off as little smoke as possible. They should burn not less than 15 minutes with a length of about 18" exclusive of handle and a diameter of ⅞".

The standard formula of 40 years ago was: for

RED.

Nitrate strontia	16
Potass: chlorate	8
Shellac,	3

however, this burns somewhat fircely and is rather expensive. A better mixing is:

Nitrate strontia	14
Potass: chlorate	4
Ground asphaltum gum	3

or

Strontium nitrate	40
Potassium chlorate	8
Red gum,	7½

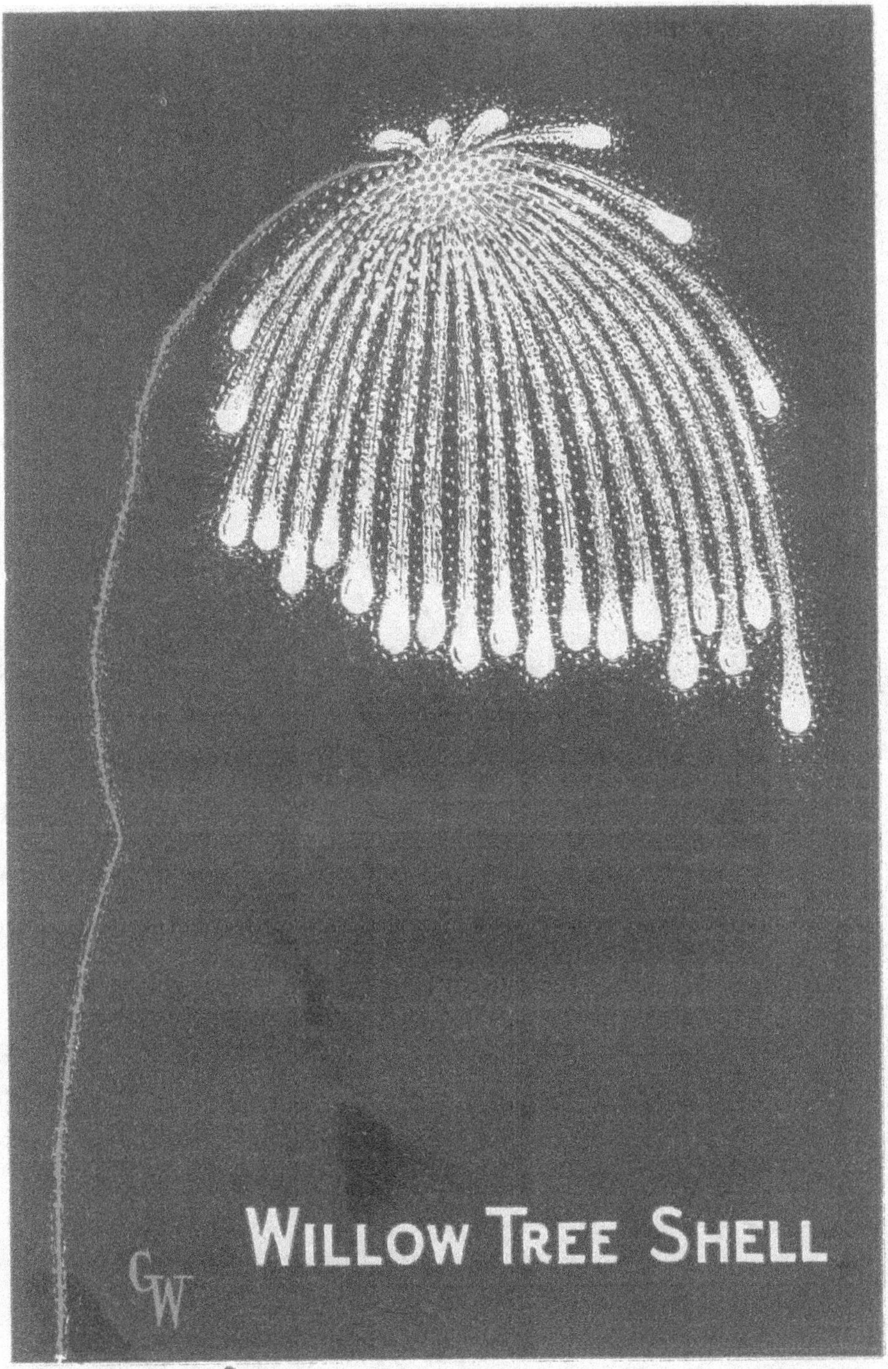
WILLOW TREE SHELL
CW

This will burn 17 minutes in an 18" torch. The latest formula, giving exceptional results is:

Strontium nitrate	9
Potassium perchlorate	2
Sulphur, ground	2
Red gum	1

This gives a fine color, burns clean and is inexpensive.

GREEN PARADE TORCHES

Barium chlorate			5
Barium nitrate	40	30	4
Potassium chlorate	11		
Potassium perchlorate		6	
*K. D. gum	6	2	
Sulphur, ground		3	
Sal ammoniac	1		
Shellac			1
Calomel			2

*K. D. and Red gum are supplied by New York dealers in pyrotechnical supplies.

BLUE PARADE TORCHES

Potassium perchlorate	5	24	24
Paris green	2		
Copper-ammonium sulphate		6	
Copper-ammonium chloride			6
Dextrine	1		
Calomel	1		
Sugar of milk		2	
Sulphur		9	
Stearine			2
Asphaltum			1

PURPLE PARADE TORCHES

Strontium nitrate	7
Potassium perchlorate	9
Black oxid of copper	6
Calomel	3
Sulphur	5

AMBER PARADE TORCHES

Strontium nitrate	36
Sodium oxalate	8
Shellac	5
Sulphur	3
Potassium perchlorate	10

The last two torches are exceptionally beautiful and have been used very effectively in canival parades. Great care must be observed in mixing compounds containing sodium oxalate, that all the ingredients are perfectly dry, and it is best in a damp climate, to mix only on a clear day, for the reason that the least moisture is liable to cause the oxalate to decompose forming sodium nitrate or chloride which is still more deliquescent than the oxalate and the work is soon so wet that it will not burn. Even when mixed in dry weather it should be protected from dampness by parafined wrappings or otherwise.

In cutting the paper for a 15 minute parade torch ⅞" diameter and 18" long cut 35 to 40 lb. Kraft paper so it will roll with the grain 18" in length and across the grain 11¾". This will give four complete turns and cause more regular burning. Pasting the outer edge for 3 or 4 inches will be sufficient.

Capping and Matching Parade Torches.

A good method of doing this is as follows:

Cut some cotton cloth into pieces about 2 inches square. Cover them with paste and bend them securely over the tops of the torches as shown in sketch (Fig. 16a). When they have dried punch a hole about 1 inch deep through the cloth and into the top of torch, with an awl about ⅛" diameter, into which insert the match. Then make

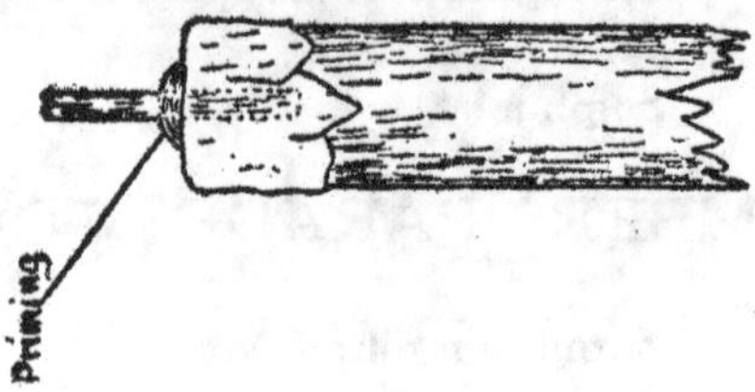

Fig. 16a.

up some thin priming of gunpowder, gum water and a little alcohol. Place this in a squirt oil can with a large opening in its spout and, shaking frequently to prevent it from separating, press out a drop or two at the point where the match enters top of torch. If this is properly done it will secure the match in place and cause the torch to ignite freely.

ALUMINUM TORCHES.

This beautiful piece of pyrotechny was first introduced in parades by the author, with sensational results, about 35 years ago. A row of 12 men was placed at the head of the line of march and with these all burning aluminum torches simultaneously there was produced the effect of an oncoming avalanche of fire. For this torch a case ½" diameter and 16" long is used with a round wooden handle 6" long. They are rammed and matched much as other parade torches and a good formula is:

Potassium perchlorate	13
Fine aluminum powder	6
Flake aluminum	5
Dextrine or licopodium	1

A beautiful modification of this is the

RED and ALUMINUM TORCH.

These should be ⅞" diameter 18" long and of the following composition

Strontium nitrate	35
Potassium perchlorate	7
Shellac	4
Coarse flake aluminum	4
Lycopodium	1

another formula is:

Strontium nitrate	13
Sulphur	3
Mixed aluminum	3

Before ramming, this formula should be moistened with a solution of 1 part shellac in 16 parts of alcohol and one part of this solution used to every 36 parts of composition. As this mixture is somewhat difficult to ignite it is necessary to scoop out a little from the top of torch and replace it with starting fire as shown in (Fig. 17).

Fig. 17

STARTING FIRE.

Saltpeter	6
Sulphur flower	4
Fine charcoal	1

An aluminum torch of heretofore unheard of brilliance and giving an illumination, in the 1 inch diameter size, of what is said to be 100,000 candle power is made as follows:

Barium nitrate	38
Mixed aluminum	9
Sulphur	2
Vaseline	1

Rub the vaseline into the barium nitrate; mix sulphur and aluminum separately; then mix with barium nitrate and vaseline. A starting fire for this also is required, as follows:

Barium nitrate	4
Saltpeter	3
Sulphur	1
Shellac	1

PORT FIRES.

These are small torches ⅜" diameter 12" long, used in exhibitions for lighting other pieces of fireworks. They are rammed with rod and funnel and a good mixing is:

Meal powder	1	
Sulphur	2	4
Saltpeter	5	5
Charcoal		1

Ship Lights and Distress Signals.

"BENGOLAS"

Another form of torch is the Bengola or Blue Light used mostly by ships in signaling for pilots. They consist of a stout paper case 1½" in diameter and 4" long, 3" of which is composition and ⅓" clay at bottom; the balance being the socket into which the handle is fitted, (Fig. 18). They should be rammed quite hard; the nosing

Fig. 18.

should be of good strong paper secured around the match with twine and the match should be piped where it passes through the nosing. The finished light should then be painted with melted parafine so as to protect it against the dampness of the sea air. This is an average formula:

Saltpeter	12
Sulphur	2
Antimony sulfid	1

Distress signals are the same except that they burn red. The regulation Life Boat equipment consists of 6 or 12 enclosed in a water tight copper can. The following formula is suitable:

Potass: chlorate	5
Strontium carbonate	1½
Shellac	1
Dextrine	½

TOY BLUE LIGHTS.

These are little lights ¼" diameter and 6" long made by rolling a light case as for lances. Cut the paper 2" x 6", the 6" way running with the grain of the paper. One end should be closed as for lances. Bunch about 200 into a bundle with string, all the open ends being uppermost when the bundle is stood on end. Now make the following composition:

Saltpeter	5
Sulphur	2
Antimony sulfid	1

When thoroughly mixed place it on a large sheet of strong paper previously spread on a firm table. Set the bundle of blue light cases alongside of the composition on the paper, with the open ends up and pour a handful of composition on top of them. Shake the bundle so as to make composition fall into the cases as much as possible and repeat several times. Now with both hands raise the bundle of partly filled lights and bring it down on the table with a good blow. Repeat this several times and then again the first operation of filling them and pounding them on the table until all are well filled when the ends may be tucked in with a dull awl.

ROMAN CANDLES.

These are probably the most popular piece of fireworks made, from a sales point of view. Up to some years ago they were made entirely by hand, that is, one at a time. Then a combination rammer taking a dozen at a time was devised. And later the Candle Machine which handles six dozen was perfected. To make roman candles by hand, roll the cases as described and have a lot of stars of different colors ready. Then make some candle composition as follows;

ROMAN CANDLE COMPOSITION

Powdered saltpeter	18 lbs.
Fine powdered charcoal	11 "
Flowers of sulphur	6 "
Dextrine	1 "
Water	1 gallon

After all the ingredients are well mixed and sifted three times add the water and mix again until the whole lot is evenly dampened. Then force through a 16 mesh sieve into cloth bottomed trays and dry in the sun.

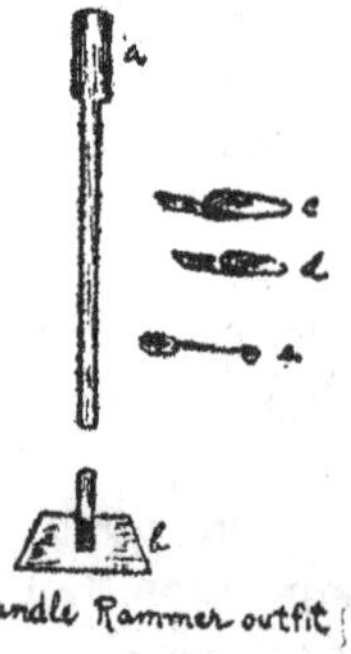

Hand Candle Rammer outfit

Fig. 19

Now provide a ramming outfit as shown in (Fig. 19) consisting of a pin block (b), a rammer (a), a composition scoop (c), a clay scoop (d) and a gunpowder scoop (e). The various parts must, of coarse, be proportioned to the size of candle it is intended to make. Say you will begin with an 8 ball. The pin of pin block must

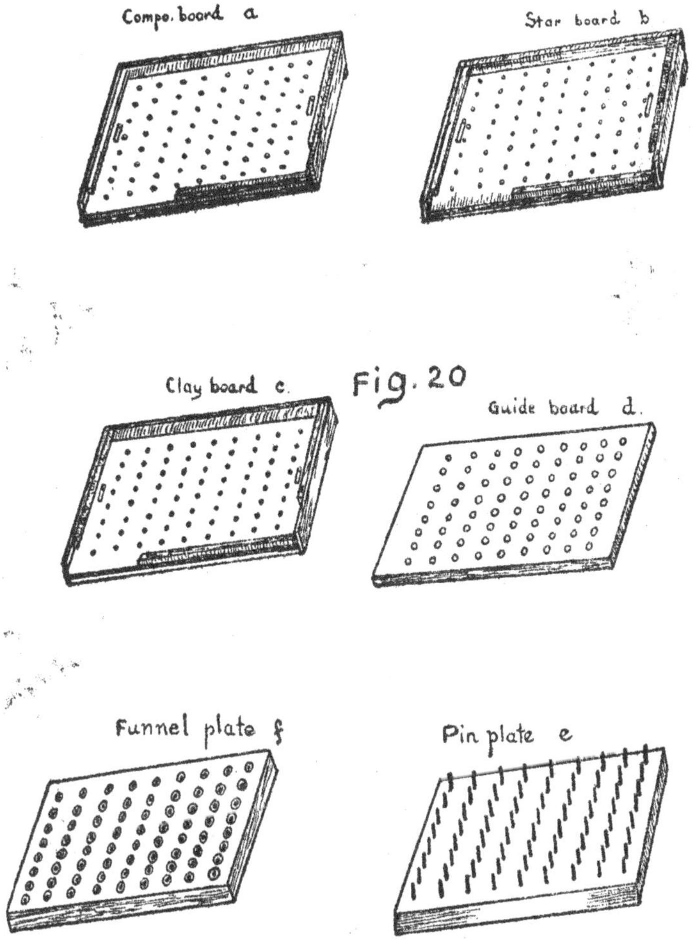

Fig. 20

be ⅜" diameter. The rammer, slightly smaller so it can pass easily up and down the candle case which also is ⅜". The clay scoop should hold a level teaspoon full of clay; the composition scoop, a heaping dessert spoonful; and the gunpowder scoop should be ¼" diameter and ¼" deep. It may be made from a .22 cal. rifle shell, if desired.

Now, place an empty case on the pin; pour in a scoop of clay and ram it firmly with a light mallet. Remove rammer; pour in a scoop of gunpowder on top of which drop a star and lastly, a scoop of candle composition. Ram with about six blows of a light mallet. Remove rammer and pour in another scoop of gunpowder; another star and another scoop of candle composition, repeating this until case is filled to within 2" of the top. Remove candle and finish as described under that head.

Hand Combination Candle Rammer.

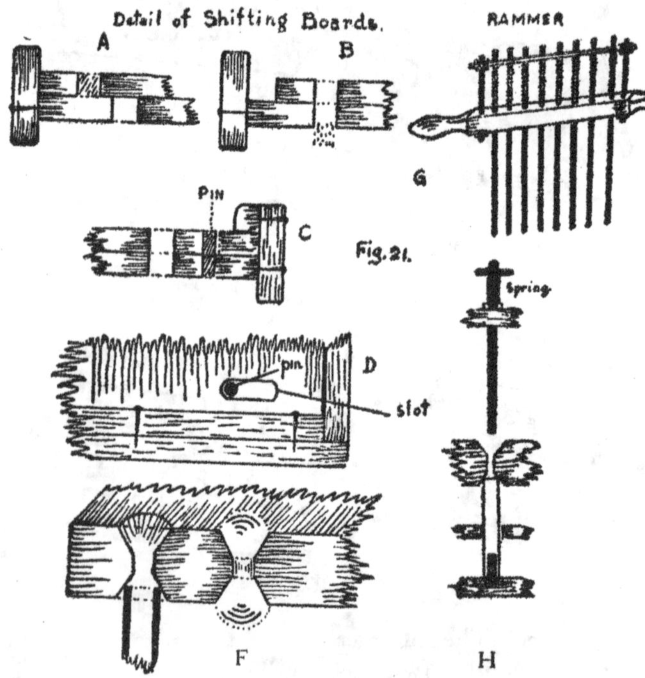

This consists of an iron pin plate (e), funnel plate, iron (f), a wooden guide board (d), three wooden shifting boards, viz: clay board (c), star board (b), and composition board (a) as well as a gun-powder box (Fig. 23 and 24) and rammer (g). The construction of shifting boards can be readily understood from detail sketch (Fig. 21). The pins shown in upper plates should be understood as being in lower plates. Otherwise slots would become clogged with composition while in use.

The holes in upper board are of a size to contain just sufficient composition clay etc. for one charge (A). This board slides a distance of about ¾" controlled by pin (CD). When upper board is pushed back the holes are filled and when ready to discharge it is drawn forward so the holes are in line with the holes in lower fixed board when the contents falls through funnel into candle being rammed, (B). The gunpowder box is described under "Candle Ramming Machine" so it is unnecessary to repeat its construction. It is of course smaller than the one for large machine and made of size to correspond to pin plate etc. Finally there is the rammer (Fig. 21), (g) consisting of eight steel rods with compression springs fitted through a wooden handle bar as shown, with detail of whole at (H).

This apparatus is used for ramming one to four ball candles and can also be used for serpents and saucissons.

Place pin plate on some solid wood block or concrete base; place guide board over pins so that the holes encircle the pins fairly; slip a candle case on each pin; place funnel plate on top of assembly and raise guide board so as to make cases center nicely.

Now, fill clay board, composition board and star board. Place clay board over funnel plate so holes are in line and shift, tapping lightly so that all clay falls out through funnel plate and into candles. With rammer give 10 to 15 strong blows through each row of holes. Put on gunpowder box and draw plate until a charge enters candles. Then take star board; place as was done with clay board and shift. See that all stars have entered candles and put on composition board. When this has been discharged give about 8 to 10 blows with the rammer, not quite as hard as for the clay. Now give another charge of gunpowder, another board of stars and a second charge of candle composition (if more than 1 ball candles) and repeat until desired number of stars have been used.

CANDLE RAMMING MACHINE.

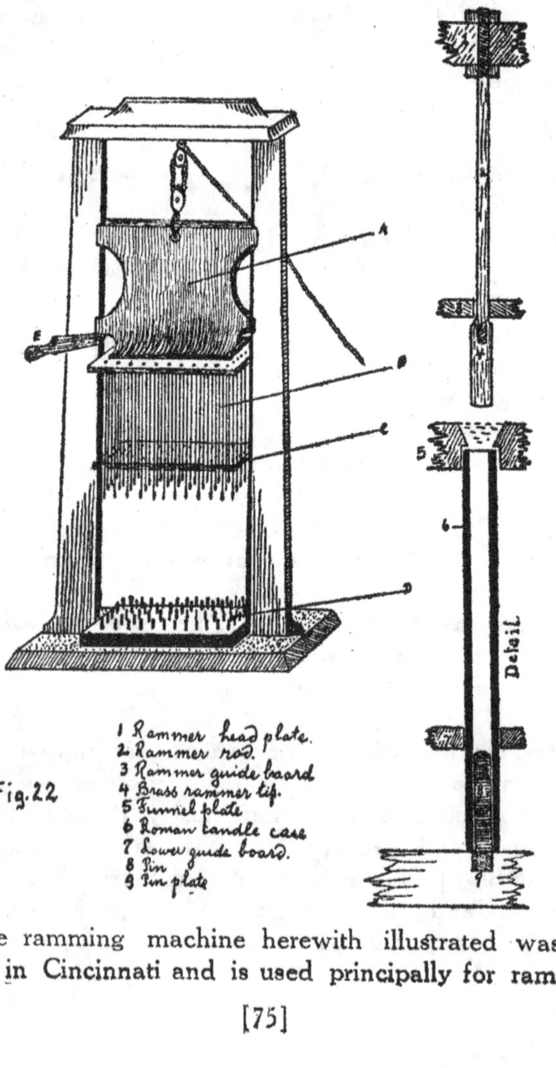

Fig. 22

The ramming machine herewith illustrated was designed in Cincinnati and is used principally for ramming roman candles from 6 ball to 30 ball but up to three ounce rockets may be rammed solid with it and the hollow center of rocket made by driving a spindle into it afterwards as will be explained later. Flower pots may also be rammed with this machine and the writer has adapted it to making 3" cannon crackers at the rate of 72 at a time. However, several sets of rammers of different lengths and thickness are required for the different sizes of candles.

The frame is of cast iron about 7 feet high; the upright sides are 1¼" thick with a V edge on inner sides upon which the head block (A) slides. The rammer assembly is fastened to head block by stud bolts. The guide board (C) is made of ½" lumber and serves to keep the rammers properly in line. This board is loose enough to slip up and down on the rammers while machine is in use. The pin plate (D) rests on base of machine and is slid into place from in front and retained by short stops in the rear. Several of these plates also are required, corresponding with the rammer assemblies as above. The pawl (E) holds the rammers up while the articles to be rammed are arranged below. When all is in place and the first charge of clay (in the case of roman candles) is in the cases an attendent pulls the rope attached to head block, which serves to disengage the pawl. The rammers are now lowered slowly until they enter the funnel plate. The rope is released and as the rammer head falls it rams the clay in the bottom of the cases. From 5 to 15 blows are usually required to ram each charge.

If composition becomes so dry that it will not pack firmly it should be dampened with a very little water. The stars should be hard and dry and free from star dust which can be sifted out by shaking stars in a coarse sieve. The floor of ramming room also should be kept free from all accumulated composition etc. to guard against accidents from friction of the shoes or otherwise.

It should here be noted that when cutting the paper for machine rammed roman candles, a thin V shaped slip should be cut from one end (a) of a sheet at side nearest the operator when being rolled. The object of this is to form a somewhat funnel shapped end to case which materially assists its easy ramming. This end must, of course, be uppermost when case is in machine. (a, b) (Fig. 16).

The funnel plate is made of cast iron one inch thick and the other dimensions being the same as head of rammer assembly. It is drilled with 72 holes in 6 rows of 12 each corresponding with the inside of the diameter of the candles to be rammed and spaced same as the rods in rammer head. These holes are countersunk on

Fig 16

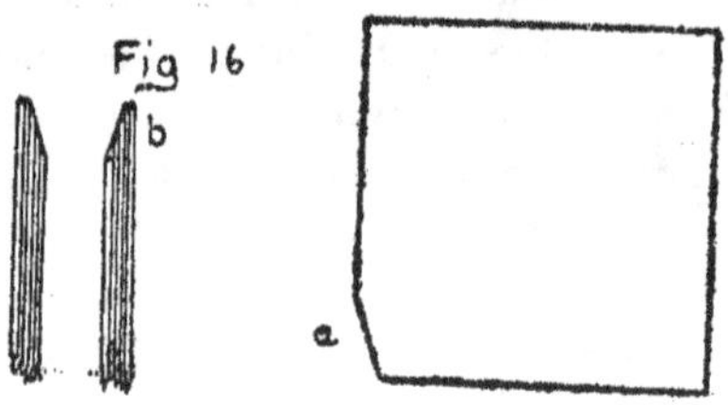

upper side of plate, to a depth of one third thickness of plate so as to give them the shape of a funnel while the under side is counter bored somewhat larger than the outside of the candles to be rammed, which slip into these recesses and thus are held in place while machine is being operated.

This funnel plate is supported in the ramming machine by an adjustable frame attached to sides of machine, which permits it to be moved up and down as required to fit the various lengths of cases to be rammed. This frame is not shown in drawing of machine.

POWDER BOX.

The powder box (Fig. 23 & 24) is made of brass

3/16" thick and its construction will be readily understood from sketch. The bottom consists of three brass plates, each ⅛" thick, drilled with 3/16" holes spaced at same distance as those in funnel plate. The holes in upper and middle plates are ½" nearer the rear of the box than the holes in the bottom plate. The upper and lower plates are fixed but the middle plate moves forward and backward ½". When it is pushed back the holes in it and the top plate are in line, so when the box is charged with rifle powder, the holes in middle plate become filled. When the center plate is drawn forward the holes in it and these in the bottom plate come into line and the little powder charge in each hole falls out into the roman candle below it. Detail at (Fig. 24) (b).

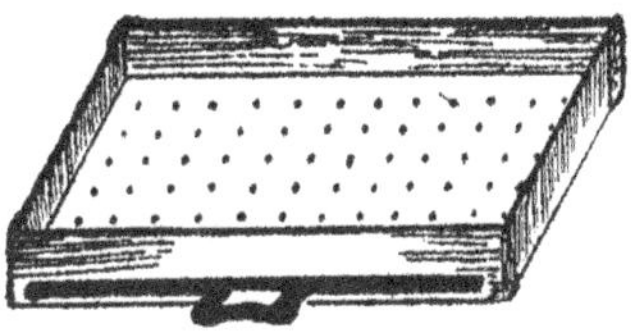

Fig. 23.

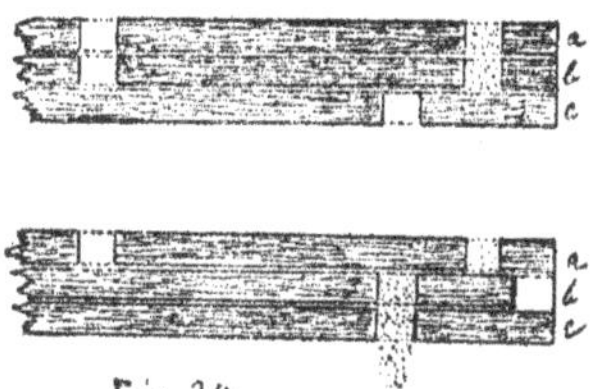

Fig. 24

To facilitate the use of this box it is placed on the adjustable stand (Fig. 25) whereon it can be raised to the desired height for the work in hand. This stand is made of light lumber and preferably on rollers so it may be moved into position and out of the way, as desired between charges.

The shifting boards follow the same principle as illustrated in hand ramming machine relative to arrangement of holes etc. but are of a size to fit other parts of big machine. Boards of different thickness must be provided so as to hold the required amount of composition for the different sizes of candles rammed. The holes in star board should be slightly larger than the stars so as to permit them to fall through easily when in use.

The stars for roman candles should be somewhat longer than their diameter as this makes them easier to fall into place when filling a shift board.

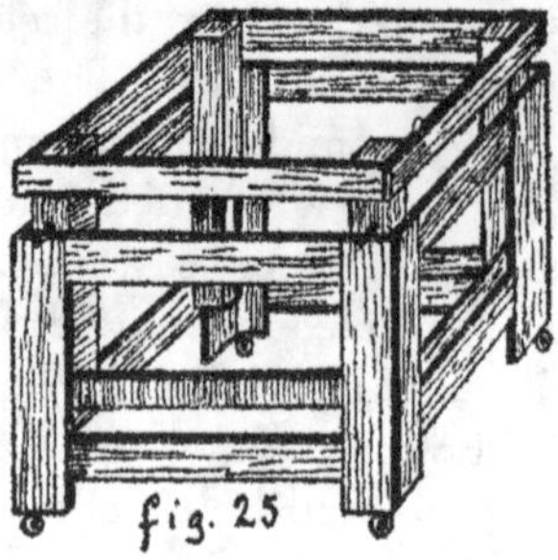

fig. 25

To fill these boards, a scoop of composition or a handful of stars is thrown on top of it; the board is shaken until the holes are evenly filled and the surplus allowed to slide off into the composition or star tray. By using boys to keep the extra pin plates and shifting boards filled as quickly as needed, and others to remove the loaded candles, a very large number of them can be loaded in a day, by one machine. A pin plate of candle cases is slipped on to the base of the machine; the funnel plate is lowered on top of it; the guide board is raised, causing the ends of cases to enter the funnel plate which is fastened in place by set screws or thumb bolts on sides of frame. The rammer head is allowed to decend sufficiently to see that all is clear. It is drawn back up into place and a shifting board of clay slipped over and its contents discharged into the candles, a slight jar being given to assure all holes of having emptied. The rammer head is now dropped some 10 or 12 times to set the clay, and withdrawn to its original position. The powder box is now slid across funnel board and by pulling handle of center plate a charge of gunpowder enters the candles. After removing powder box a board of stars is shifted into the funnel plate. Care must now be used to see that all stars have slipped through funnel plate into candles. Now a board of composition is discharged the same way and the whole rammed with about 8 to

10 blows. This operation is repeated as often as the size of candle requires. When last charge of composition has been rammed the pin plate of candles is removed, unloaded and refilled with empty cases while another pin plate of empty cases has been slipped into its place in machine.

BATTERIES.

A very effective piece of fireworks (Fig. 26) easily

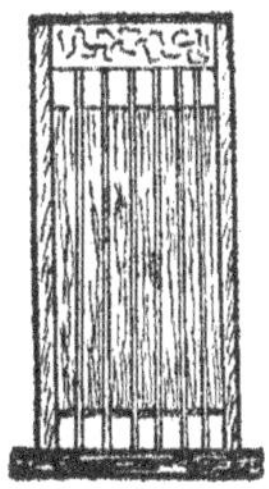

Fig. 26

made by taking a wood box about two inches longer than the candles to be used and filling it with about three dozen 8 ball or 10 ball roman candles. The space above the candles in the box is to be filled with a few scraps of match, one piece allowed to hang over the side and a piece of cardboard nailed over for a top.

BOMBETTE FOUNTAINS.

These are an effective combination of candles and floral shells packed in a large box as shown in (Fig. 27).

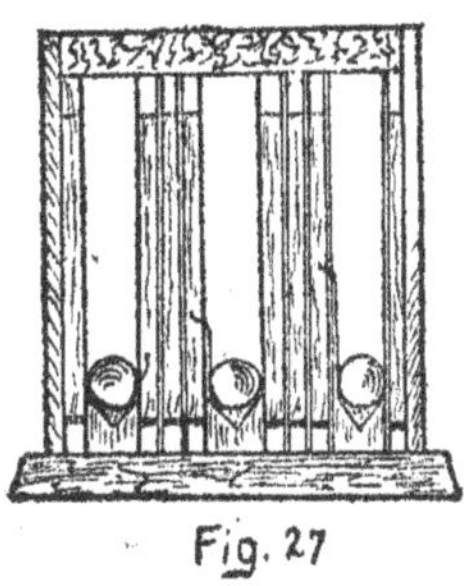

Fig. 27

All are lighted at once by scraps of match in the top, but the floral shells are matched so as to fire alternately, one at a time as shown, during the burning of the candles.

Another interesting use for roman candles is in the so called union battery (Fig. 28) which consisted originally

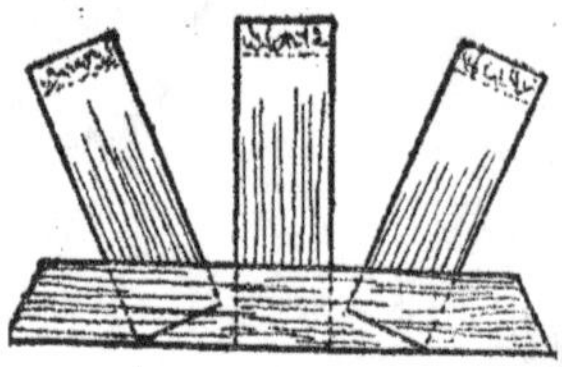

Fig. 28

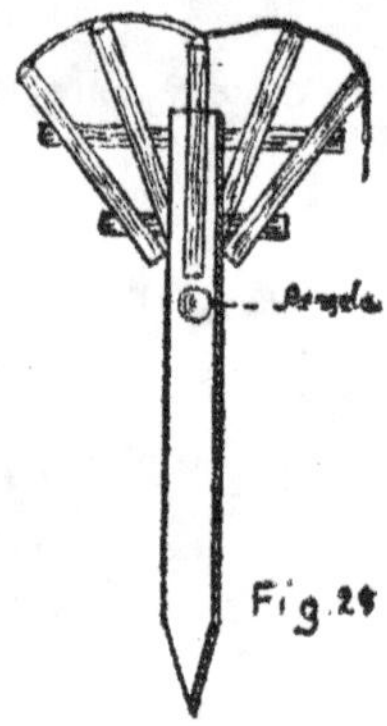

Fig. 29

of one battery each of candles containing white, red and blue stars. It is however, now used effectively with candles of varigated stars.

BENGOLA BATTERY.

Fireworks displays are often started with a row of vari-colored lights or bengolas set about 25 feet apart in front of the set pieces. When these are supported by a fan of candles or gerbs a very effective display is produced. The bengolas are lighted first and when they have burned half way, the candles or gerbs are lighted. (Fig. 29)

SKY ROCKETS.

Next to roman candles, these are perhaps the most popular article of the pyrotechnical craft and on good authority, seem to have antedated the candle. So much has been written about sky rockets that any general description would be superfluous. The French, in particular, have left a most detailed history, sometimes amusing in view of present day conditions, regarding its manufacture.

Fig. 30

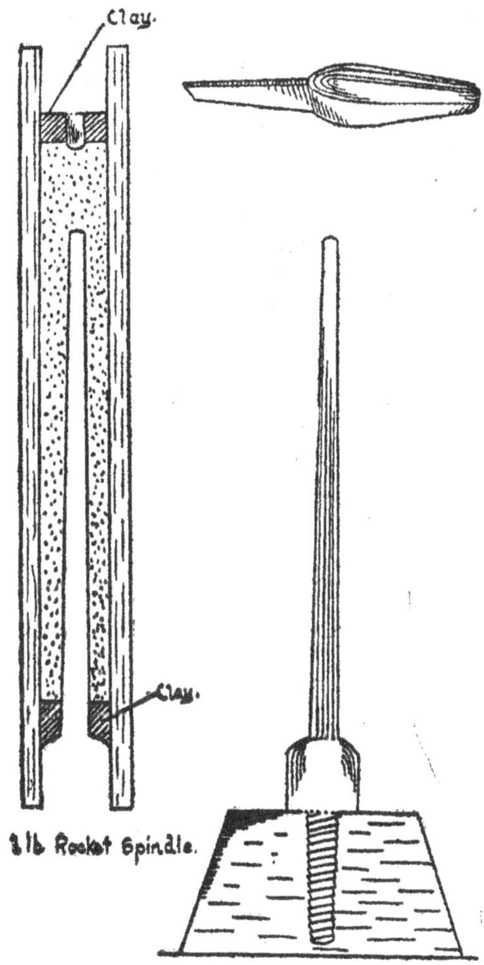

Suffice to say that the rocket consists of a tube of

paper, rammed with suitable composition, its lower end choked to about one third the diameter of its bore and a hollow center extending upward though the composition to about ¾ inch of the top. A stick attached to the tube serves to balance it while ascending. Broadly the composition of a rocket, that is the portion of it which burns while it is ascending, should be seven times its diameter, in length. Six sevenths is pierced through the center while one seventh is solid and acts as the fuse to communicate the fire to the heading when rocket

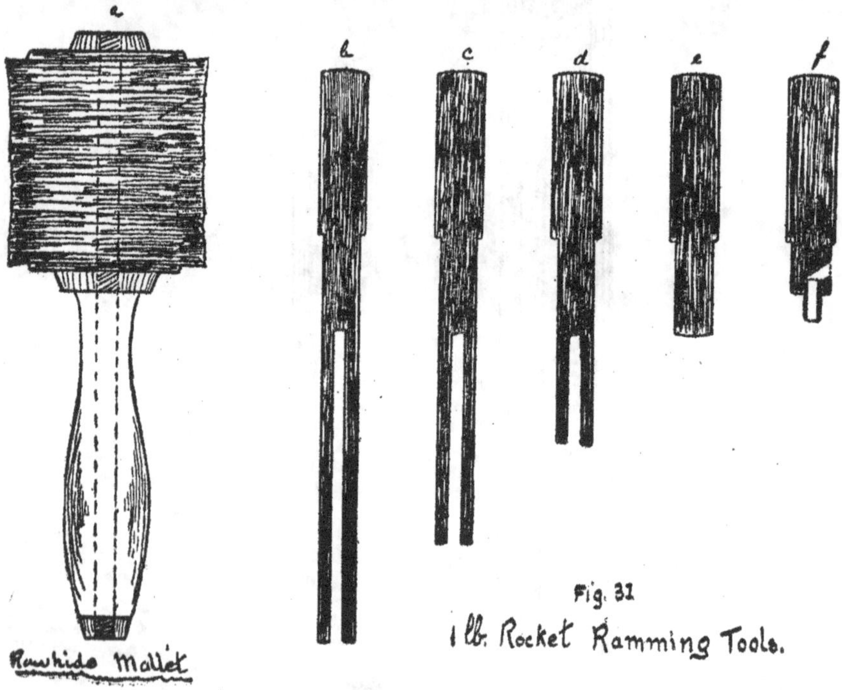

Fig. 31
1 lb. Rocket Ramming Tools.

reaches the highest point of its flight. The tube is made of good strong paper. preferably three turns of hardware on the inside with four or more turns of good strawboard on the outside, but a good rocket case can be made of heavy rag or building paper if properly rolled with good paste. Choking the case and ramming in mold has been practically discontinued.

An average model for a 1 lb. rocket is given. in (Fig. 30) with a corresponding set of ramming tools in (Fig. 31). The spindle is one half actual size while the ramming tools are one third actual size.

Good rockets should be uniform, all those of one caliber ascending to approximately the same height and exploding at about the same time. Particularly is this desirable in bouquets or flights of 100 or more fired simultaneously, else a straggling effect is produced.

Most rockets larger that 3 oz. are rammed singly or by gang rammers as shown in sketch (Fig. 32) but hydraulic rammers are also in use. See Military Pyrotechnics, H. B. Faber, Vol. 2 pp. 39.

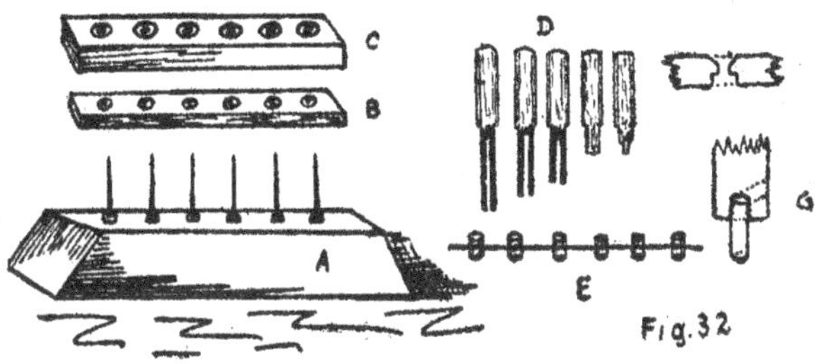

Fig. 32

The gang rammer is quite efficient and one man can get out a large number of rockets in a day with it. A shows the spindle block; B is the guide board for assisting to get the ends of cases into funnel piece C. D shows a set of rammers while E is the set of scoops for charging the entire six cases at once. It is easily made by cutting brass shotgun shells in half and soldering them to brass rod as shown. Details of funnel piece and hollow pin rammer used in setting top clay charge are shown at F and G respectively.

1 oz. to 3 oz. rockets are rammed solid on the candle machine or otherwise and the hollow center is made by driving a steel spindle into them afterwards. These must

have their lower ends choked as explained on page 41. An efficient way of doing this is to get a mortising machine and replace the chisel with the spindle as above. A V shaped block is set on table of machine, in such a position that when a rocket is placed on it, it will be in just the right position for the spindle to enter it. A step on the pedal of mortising machine will force the spindle into the rocket and make the necessary hollow center.

To ram rockets from 4 to 8 oz. singly the case is slipped on the spindle illustrated under sky rockets; a scoop of clay is shaken in and rammed by about eight good blows of the mallet on the longest rammer. Then a scoopful of composition is rammed with about eight lighter blows. This is repeated until the case is filled to within about 1 inch of the top, shifting rammers as it becomes possible to use shorter ones. Now the final charge of clay is put in and the hollow pin rammer used. This sets the clay while leaving an opening for the fire to reach the heading. Care must be taken to see that the hollow tube just pierces the clay. If it does not go through, the heading will fail to fire; if it goes through too far, heading will fire prematurely or rocket is liable to blow through before rising.

The following are good compositions for rockets of different sizes:

	1 to 3 oz.	4 to 8 oz.	1 to 3 lb.	4 to 8 lb.
Saltpeter	18	16	16	18
Mixed coal	10	9	12	12
Sulphur	3	4	3	3

If rockets burst add more coal; if they ascend too slowly add more saltpeter. For the smaller sizes use fine coal; for the larger ones use coarser coal in proportion to their diameters. 4 lb. to 8 lb. rockets use granulated saltpeter.

All rockets larger than 3 oz. are provided with a cone to contain the heading. These are made as follows:

SKY ROCKET CONES

Turn out a cone former on the lathe, of a shape somewhat as shown in Fig. 33. Cut some stiff paper to the shape of one third of a circle, the radius of which for a 1 lb. rocket should be 3 inches. Lay it on the table before you with the round side toward the right.

Fig. 33

Paste the straight edge farthest from you and place the former on it with the point toward the left and about ⅜" from the point on the paper where the two straight edges meet. Now roll it around the former commencing with the unpasted edge. When finished slip off former to dry.

HEADING ROCKETS

Prepare a board with holes through it about 1¼" in diameter and raised from the table about 3" as shown in Fig. 34. Place in these holes a number of the cones, point down, and fill them about half full of stars, gold rain etc. Also a little meal powder and charcoal or candle composition. Apply gum to the upper edge of a rocket

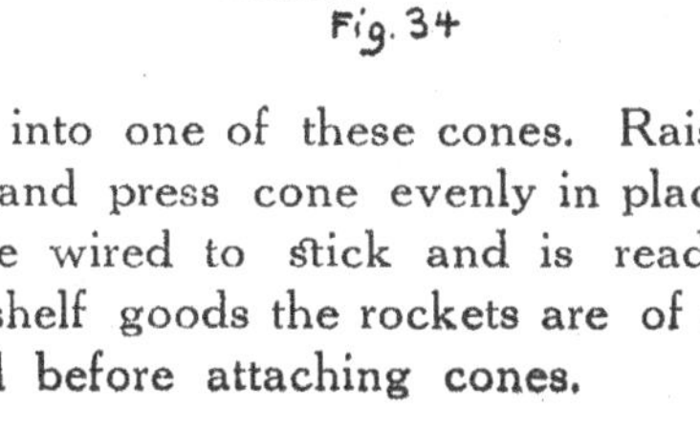

Fig. 34

and stick it into one of these cones. Raise carefully out of the hole and press cone evenly in place. The rocket may now be wired to stick and is ready for use. In the case of shelf goods the rockets are of course papered and matched before attaching cones.

SHORT STICK ROCKETS.

These are the same as long sticks except that a stick only ⅓ the regular length is used, on the bottom of which a wing or tab of cardboard is attached (Fig. 35a). Cut a piece about 3" long, 1½" wide at one end and ½" wide at the other. Smear a little dextrine on one end

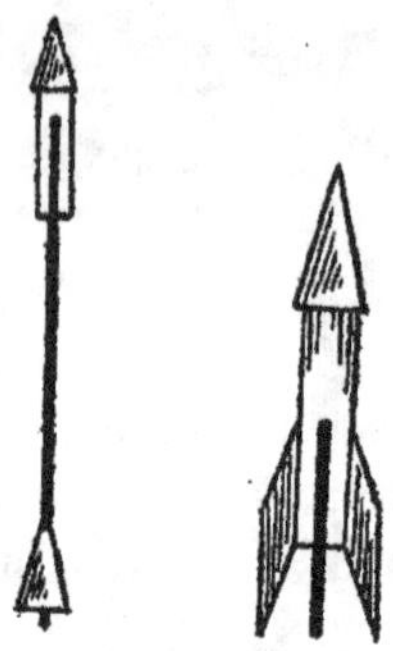

Fig. 35a.

of stick, place the tab on it, large end down and drive a 2 oz. tack through it in the middle. When dry it is ready for use. These rockets are much easier to carry about but require more care in firing to get them started straight.

The one illustrated at right (Fig. 35a) has no stick at all, only four wings, upon the ends of which it rests when lighted.

When the bottoms of brass rammers become worn from use, they may be reconditioned by battering them until they are again full sized on the ends.

There are a great many of so called fancy rockets in which the heading is not confined to a simple burst of stars etc. but is supplemented by many other beautiful effects, some of which will be described here while the ingenuity of the pyrotechnist must be relied on for others.

WILLOW TREE ROCKETS.

These are made by filling a large rocket head with pieces of Japanese Star and a weak bursting charge. If the bursting is too strong many pieces will fail to light.

Prize Cometic, or Shooting Star Rockets.

These are prepared by placing 4 or 5 four ounce rockets, without sticks, in the head of 6 lb. rocket besides a handful of box stars. A few #1 stars are also placed in the top of each of the 4 oz. rockets with a pinch of grain powder, and well capped.

GOLDEN CLOUD ROCKETS.

For these the rocket head is filled half full each of gold rain and aluminum stars. The weight of the contents of a rocket head must be proportioned to the size of the rocket. A heading of heavy stars must be smaller than one of lighter materials.

BOOM ROCKETS.

Have one or more small maroons in the head besides a few stars.

ELECTRIC SHOWER ROCKETS.

Made by filling a small head with electric spreader or granite stars. As these are very heavy only a small quantity can be used.

BOMBSHELL ROCKETS.

These have a small shell with very short fuse fastened to top of rocket, with a few stars in the head of the rocket itself, which burn before the shell bursts.

WHISTLING or CALLIOPE ROCKETS.

The head of this rocket is filled with whistles, made as described under that caption. In addition, a few colored stars are added.

LIQUID FIRE ROCKETS.

These are one of the most beautiful pyrotechnical effects known to the art. Take a 3 lb. rocket and fill the space above the clay with grain powder. Cover this with a circular piece of perforated paper secured by a strip of tissue paper. Roll on a head of about three turns of strong manilla paper, only pasted on the edge, about 6" long. Now procure some sticks of phosphorus and cut them under water with a chisel into pieces about ¾" long. Get some ¼ lb. tin cans, punch a number of holes in the bottoms of them and fill with the pieces of phosphorus, conducting the entire operation under water. When ready to fire the rockets remove one of the cans from the water, allow to drain for a few seconds, empty contents into one of the rocket heads tuck in and fire at once. Great caution must be observed owing to the dangerous nature of the phosphorus.

PARACHUTE ROCKETS.

To successfully launch a parachute from a shell or rocket requires the greatest care and skill, besides patient attention to every detail or the light fabric will either fail to unfold or be torn or burned in its exit from the tube in which it is placed. To begin with procure some very light Japanese tissue paper; cut into spuares about 18 inches each way and rub thoroughly with powdered soapstone.* Cut four pieces of stout linen twine or shoemakers thread about 18" long. Twist the corners of the tissue squares a little and tie a thread to each. Draw the other four ends of threads together and tie them in a knot. The parachute is now ready to fold. In one hand take the knot where the four strings meet and in the other take the top of the parachute by the center. Draw the hands apart until the paper folds together and lay on the table in front of you. Straighten out the four folds, two each way, and fold them again laterally toward

---

*Ready made parachutes may now be purchased from stock.

the center about five or six times like the bellows of an accordeon until the pile is about 1 inch wide. Now roll this up lightly beginning at the small end or tip until you come to the strings, then wind the four strings, also lightly, around the bundled parachute until it will just about fit the rocket head for which it is intended.

For making the light ram a short case ¾" diameter and 1" long with box star composition. Prime one end and stop the other with clay. Over clayed end glue a cardboard disc slightly smaller than inside diameter of rocket head, having first passed a wire through the case under the disc so as to form a loop on exposed size as shown, (Fig. 35). Pass about 18 inches of stout linen

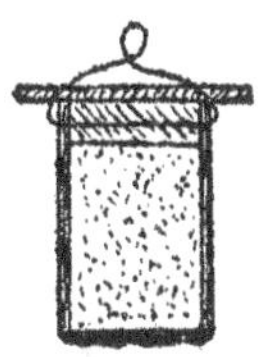

Fig. 35

twine through the wire loop and tie the other end to the knot on the parachute where the four strings come together. Roll a piece of naked match about 18" long, into a bunch and place it in the bottom of the rocket head for a blowing charge. On top of this drop the primed end of the parachute light and over it place a small wad of cotton waste and a little cotton hulls or bran. Now slip in the parachute around which the strings have been lightly wound. Fill all around parachute with bran and secure the top of rocket head very lightly so the parachute will be thrown out when discharged, with the least possible effort.

CHAIN ROCKETS, (Caterpilars).

If you have succeeded with parachute rockets you may now attempt this modification of the above which is infinitely more difficult but their great beauty compensates for the trouble required in their preparation. A parachute several times larger than the first described is made in substantially the same manner but preferably octagonal with the separate pieces sewed together. Instead of one light, a dozen or so of different colored lights are attached to it. This is called the chain. and to launch it successfully from either rocket or shell is about as difficult a proposition as the pyrotechnist is called upon to execute.

For the lights composing the chain ordinary lances may be used. To a 4 lb. rocket take 12. Procure a strong linen twine about 18 feet long. To this attach the lances at intervals of about 1½ feet, by taking two half hitches around the bottom ends. (It is best to make special lances for this purpose, filling the first ⅜ inch at the bottoms with clay). When all are fastened, tie one end of chain to parachute and at the other begin to wind up the slack between the lances. Wind each lance with the slack between it and the next one to it, winding as smoothly as possible without lapping the twine anywhere. As each one is wound lay it against the other one before it until the 12 are in a round bundle. Then take a few turns around the entire bunch on upper end, so as to hold it together. At the bottom end of bunch take two turns of light cord not more than ¼ inch from the end. This is to hold the lot together until the lights all take fire when this cord burns off and chain unwinds in the air. A cardboard wad fitting easily in the rocket head and with a hole through its center is placed on top of the primed ends of the bunch of lances and a piece of match passed through the hole in same so as to touch them. This may be fastened in place with a small tack or two.

Now prepare the rocket head for the reception of the chain, as directed for parachute rockets, by placing about 2 feet of naked match in bottom of head for the blowing charge. Slip the bunch of lances on top of this with another paper disc, through which line runs, over it. Put in a good wad of cotton waste, then the parachute carefully folded as described and pack with bran. Now cap the rocket head as lightly as possible and if all directions have been carefully followed, the chain will be likely to come out successfully. A few trials, however, are generally necessary. Sometimes four light sticks are inserted in rocket head alongside of the parachute the lower ends resting on a stout wad under the bunch of lances and the other ends against top disc over parachute. This is to keep parachute from being injured while being expelled by blowing charge.

BOUQUET or FLIGHT of ROCKETS.

These are made by firing a hundred or more rockets at once from a specially prepared box. Take three boards of ½" lumber, 12" wide and 4 feet long; clamp two of them together and with a ½" bit bore 5 rows of holes 2 inches apart and beginning 2 inches from the sides

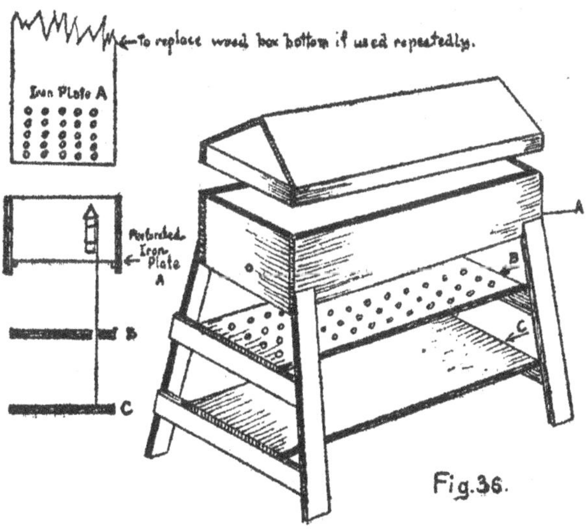

Fig. 36.

and ends. This will make 100 holes through the boards (B). Now make a box, the bottom of which is made of one of the boards with holes through it as shown in

(Fig. 36) (A). Attach four legs to the box, about 4½ feet long. At 1½ feet from the bottom secure the other board with the holes in it, (B) so that a rocket passed through a hole in the box bottom may be steadied by passing through corresponding hole in lower one. Fit the third board (C) in the legs also, about 6 inches from the ground to make a resting place for the rocket sticks and so as to hold the bottoms of the rockets in box, about one inch above bottom board (A) of box. This is to permit the fire to reach all the rockets instantly when flight is lighted.

Flight rockets used this way need not be matched; only primed and a little loose grain powder thrown on bottom, inside of box and a piece of match passed through a hole in side to fire it from is all that is required. If a top, covered with canvas, is fitted to flight box, same may safely be left in the rain until required. Some pyrotechnists make flights by stringing rockets in a row on slats provided with nails to hold them apart but the effect in much inferior.

ROCKET STAND.

The best method of firing sky rockets is from a wooden trough constructed of two light boards, ½" thick,

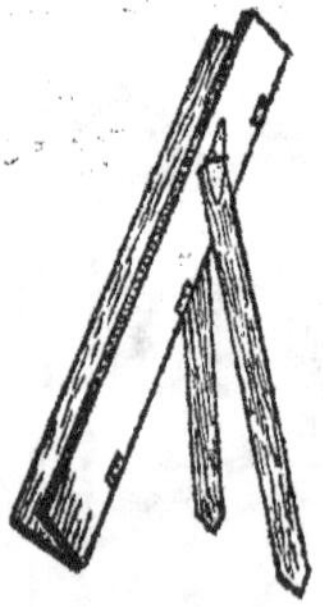

Fig. 37

4" wide and 6' long. These are nailed together so as to form a gutter and supported by two legs. If the boards and legs are hinged as shown in Fig. 37 the trough may be folded and easily carried about.

TOURBILLIONS.

(Geysers, Whirlwinds, Table Rockets)

This is a modification of the sky rocket and ascends to a height of about 100 feet, in a spiral manner and without a stick. They are made by ramming a 3 lb. rocket case with one of the following mixtures:

Saltpeter	8	5
Meal powder	7	12
Charcoal	2	3
Sulphur	2	3
Steel filings	3	

Both ends of the case are stopped tight with clay. Four holes are bored in it, ¼" diameter. Two are bored into the bottom, 3" apart or 1½" each way from the center and one hole on each side, 1" from end and opposite to each other as shown in sketch (Fig. 38). A piece of

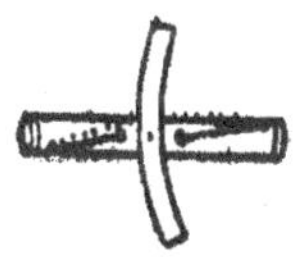

Fig. 38

curved stick, as long as the case is nailed to the bottom of case, concave side down exactly in the center and at right angles with same. The holes are primed and while still wet it is matched by tacking a piece of quick match to one of the bottom holes, passing it to the nearest end hole; then over the top to the other end hole and finally to the other bottom hole. A small hole is now made in the match pipe as it passes over the top of case, just in the center, into which a short piece of naked match is slipped for lighting. To fire a tourbillion it should be laid on a wide board or smooth surface stick down and lighted with a long portfire.

Small tourbillions are sometimes made by boring only two holes in the under side of case, at an angle of 45° from the perpendicular, but those with four holes, especially in the larger sizes are safer and more likely to function. They may also be matched by using naked match all around and afterwards covering the whole tourbillion with tissue paper pasted and pressed closely to same. Large tourbillions are sometimes further beautified by placing a few stars in the ends of the case, outside of the clay, boring a small hole through same and securing over top with strong paper and a wad. A little meal powder is put in with the stars and when the tourbillion reaches its height, these are thrown out with fine effect.

FLYING PIGEONS.

This amusing piece of fireworks is easily made in its simplest form by securing two rockets with their openings pointing in opposite directions, to an empty case as shown in Fig. 39 (a). The rear end of one is connected

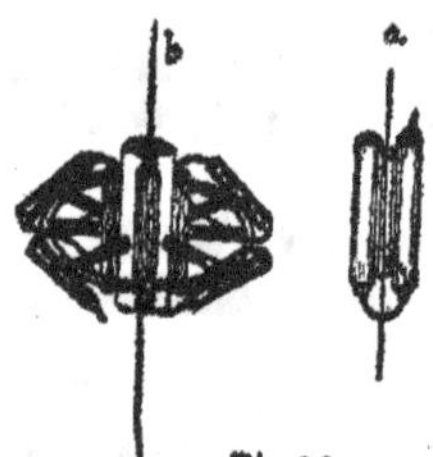

Fig. 39

by a piece of match to the front of the other. A piece of thin rope or telegraph wire is stretched between two posts about 20 feet high and 300 feet apart. One end of wire is previously slipped through the empty case forming the middle of pigeon. On lighting the first rocket the pigeon will run along the line until the other rocket lights when it will return to the starting point.

A more elaborate form of pigeon (b) is made by procuring a frame as shown. This consists of a vertical wheel frame with a heavy slotted hub. A row of 4 wheel cases are fastened to the rim and four 1 lb. rockets are secured to the long slots in hub, two pointing each way. The pigeon starts with one of the wheel cases, the rear end of which is connected to one of the rockets. This in turn is matched to the second wheel case and that to the next rocket, pointing in the opposite direction, and so on to the last rocket.

English Cracker or Grasshopper.

Cut some good 20 lb. 24" x 36" manilla or kraft paper into strips 4" wide and 12" long. If cut with the grain of the paper as it should be this will give 18 cuts from one sheet. Roll them into short tubes as directed for match pipes, getting the opening at one end, somewhat larger than that at the other. This may be done by rolling a V shaped strip of paper on one end of rod. When a quantity of these tubes have been rolled close the smallest end by twisting or folding it over. Dry them in the shade and put about 12 dozen in a bundle, all the open ends one way. Stand the bundle on a large sheet of paper with the open ends up and pour FFF rifle powder on top of it until all the tubes are full. Jolt bundle occasionally to be sure none are only partly filled. Then draw them out, closing the top end as you did the bottom and wrap them all in a wet towel, setting aside in a damp cool place for several hours. A good way is to take a long cloth, wet it well, spread the loaded pipes loosely on it and roll it up so that each pipe will touch part of the wet cloth as they should be moistened through but not wet before proceeding further. When this condition has been reached (on which the whole success of the operation depends) run them through a clothes wringer or other roller so that they will be somewhat flattened. The exact amount of flattening can only be found by experiment.

GW
LIQUID FIRE ROCKET

Now take a piece of wood, say 1" thick and 4" wide by 18" long. Notch out a piece as shown in Fig. 40 1¼" wide and 6" deep and procure a dozen pieces of stiff wire 4" long. Lay the lower ends of a half dozen of the damp pipes across the bottom of the notched

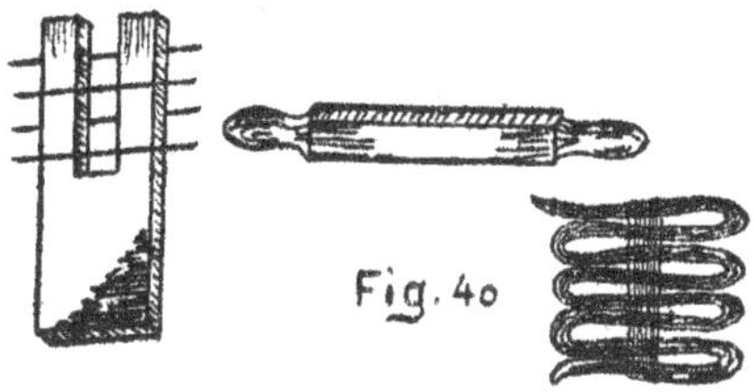

Fig. 40

board which has been fastened in an upright position to a bench. On top of these and against one side of the board lay a wire and bend the pipes across over it until they now point in the opposite direction. Lay another wire as before but on opposite side and repeat the operation until the entire length of the pipes have been folded up. Then take a bar of wood shaped as shown in sketch and, holding one end in each hand press the folded pipes down as hard as possible so as to have the turns well formed. Now lift out the folded bunch, wires and all. Remove wires, fold bended pipes, one by one in the hand and with linen shoemakers thread secure them by wrapping half a dozen turns around the folded pipe and finally pass a few turns between the folds. Strip off one end so the powder is exposed and prime it with a little wet powder or match it; or the end may be twisted up with touch paper, made by coating unglazed paper with a solution of saltpeter, before folding. When dry crackers are finished.

In this country a cruder form of cracker is made by taking suitable lengths of covered match, damping and folding it like the crackers described, tying and leaving a short piece of match protruding for lighting it.

PIN WHEELS.

Fig. 41

For making these proceed just as described for English Crackers except using the following composition instead of gunpowder for loading them:

Meal powder		10	8	2
Fine grain powder	8	5	8	
Aluminum			3	
Saltpeter	14	4	16	1
Steel filings	6	6		
Sulphur	4	1	3	1
Charcoal	3	1	8	

When they are dampened and rolled out punch out a lot of round pieces of #60 strawboard, with a hole through their center. Then get a piece of brass, the same size as the cardboard centers and fasten it to the work table. Lay one of the centers on this brass plate and taking a filled pinwheel tube press the smallest flat end against its edge and twisting it around disc with the right hand while left hand feeds the tube as it is being wound on, continue until all the tube is rolled around the center. The brass plate should be half as thick as the finished pin wheel so the cardboard center will be held just about in the middle of the pinwheel while it is being twisted.

Now have some boards prepared with strips of wood ¼" square, nailed on them, the same distance apart as the width of a pinwheel when it is lying down. When the wet pinwheel is twisted up as above, lift it off the brass plate and set it between two of these strips on the board so as to keep it from untwisting and with a brush put a drop of glue across the pipes and onto the center disc, at four equidistant points. When they have dried they may be removed from the boards and are ready for use. (Fig. 41).

SERPENTS or NIGGER CHASERS.
(Squibs)

These are light strong cases, 3" to 5" long, crimped at one end and charged with a sharp composition, strong enough to cause them to run around on the ground or in the air while burning. They may be made from #140 strawboard, heavy manilla or rag paper and crimped while still wet. (Fig. 42).

Fig. 42

They may be rammed singly with rod and funnel or in batches of 72 at a time with the hand combination rammer, q. v. Alternate compositions are:

Meal powder	3	3
Saltpeter	2	5
Sulphur	1	1
Mixed coal	1½	¾
FFF grain powder	4	3

SAUCISSONS.

These are very similar to serpents but somewhat larger and always end with a report. The usual length is 3½" with a diameter of ⅜" to ½", rolled and crimped like serpents though with a heavier case.

Ram with,	Meal powder	4
	Saltpeter	2
	Fine coal	1½
	Sulphur	1

For exhibitions, about three dozen of these are put in a paper bag with three onces of blowing charge composed of half meal powder and half grain powder. A piece of match a yard long, bared for an inch or two is stuck into the mouth of the bag and tightly secured with a string. When ready for use it is loaded into a mortar and match ignited.

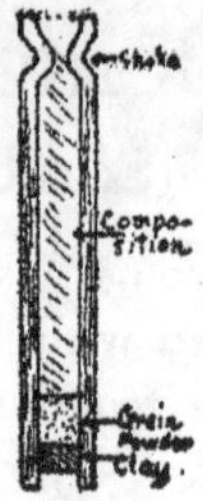

Saucisson

Fig. 43

For stock work a paper mortar is made by rolling six or eight thicknesses of heavy strawboard 12" wide around a former 2½" diameter. A wooden bottom is fitted and a mine bag made as described under MINES. The saucissons are placed in same with blowing charge, around a 10 ball roman candle from which the bottom clay has been omitted. This is placed in the paper mortar with a daub of glue on bottom of bag. A top is fitted as for mines and when papered and striped, is ready for the market.

MINES.

These are small paper guns from 1" to 3" in diameter in the bottoms of which are placed small bags of stars powder etc. which are fired by a mine fuse or roman candle in which the charge of clay has been omitted and replaced by one of candle composition. The bottoms are turned out of wood. The tubes are made by tightly rolling six to twelve thicknesses of strawboard, around a suitable former. Following are approximate sizes:

Fig. 44.

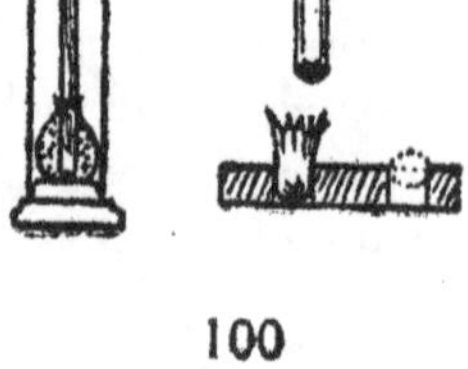

NUMBER	HEIGHT	DIAMETER	NUMBER OF STRAWBOARD
1	4"	1½"	1 sheet #140
2	4¾"	1¾"	1 " #120
3	5½"	2"	1 " #100
4	7"	2¼"	2 " #100
5	8½"	2⅜"	3 " #100
6	10"	2⅞"	4 " #100

No. 1 mines take a 1 ball mine fuse; No. 2 mines a 2 ball fuse fuse etc.

The mine bags are made by boring a number of holes into a thick board; (for #1 mines, 1¼" diameter and 1½" deep). Then make a punch with rounded edges; (for #1 mines, 1" diameter) and over this press a piece of stout paper (about 4" square) closely around end of punch and shove it into one of the holes in the board; remove punch; insert a mine fuse and around it put a half once of stars and a teaspoonful of blowing charge. Squeeze loose ends of bag around fuse and secure with a piece of string or wire. Now daub the bottom of bag with a little dextrine or glue and insert it in one of the paper guns into which a bottom has been previously glued. A top is now necessary. This is made by adjusting an ordinary washer cutter to the requisite size so as to cut a piece of strawboard with the outside diameter of the gun and center hole of the size of the mine fuse When this is slipped into place over the fuse it is secured by a square piece of paper, an inch or two larger than the top of mine and with a hole punched in the middle with a wad cutter, to fit over mine fuse. Paste and press closely about the top of mine and when dry same is ready for use. For stock work they must, of course, be papered and striped, packed and labeled. Mines of saucissons are made by substituting saucissons for the stars. (Fig. 44).

DEVIL AMONG THE TAILORS.

These are made by taking a large short mine case and filling the bag with tailed stars, serpents and English

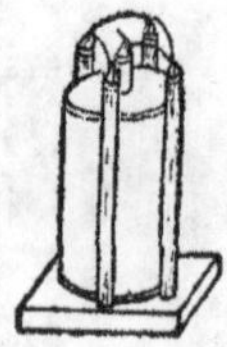

Fig. 45

crackers. Besides the central candle for firing it, four more candles, one at each corner on the outside of gun are fastened and connected so as to burn at same time. (Fig. 45).

ANGLO JAPANESE MINES.

These consist of a #6 mine case containing a bag filled with colored stars and Japanese or Willow Tree stars. Electric Spreader stars with crackers also make a handsome mine. The various effects are almost unlimited and the genius of the artificer will suggest other combinations

Fountains, Flower Pots and Gerbs.

These are all modifications of the same principle which is a paper tube or case varying from ½" diameter to 2" diameter rammed solid with one of the compositions to be given later.

FOUNTAINS.

Are usually from 1" to 1½" diameter and 12" long with a wooden point in the lower end so they can be stuck in the ground for firing. A quarter ounce of rifle powder is sometimes placed after the last charge of composition and before the clay, both in fountains and gerbs so as to have them finish with a report or "bounce".

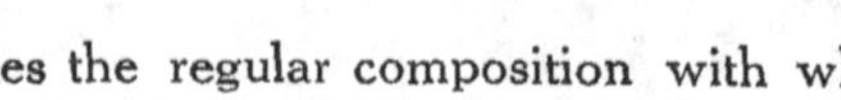

Fig. 46

Besides the regular composition with which fountains are charged, if the calibre permits, small colored stars cut to about ¼" cubes and placed between the charges when ramming, greatly increase their beauty and they are then called FLORAL FOUNTAINS or PRISMATIC FOUNTAINS. There is however, some danger in ramming stars containing chlorate of potash with compositions containing free sulphur and this may be avoided by using compositions free from chlorate, such as granite stars, copper borings etc. or perchlorate compounds.

Cascade cases are used for water falls and such designs where the fire is required to fall considerable distances to the ground. They are usually from 1½" to 2" diameter and 12" long. Where this piece is to be often repeated as at Fairs, iron tubes 2" inside diameter are sometimes used as these are stronger and can be cleaned with kerosene after using for repetition. Where Niagara Falls is shown this form of case is in general use as it saves the rolling of 200 to 300 large cases for each display.

FLOWER POTS.

Fig. 47

Small cases, choked, ½", ¾" and ⅝" diameter and from 5" to 10" long with a wooden handle in end provide a pretty piece of fireworks for use by ladies and children. When properly made they are perfectly safe to fire from the hand but this fact should be assured by first firing a few by sticking them in the ground, to see that the charge is not sufficiently strong to burst the case. The lampblack in these produces a peculiar effect not entirely understood.

It might be well to mention that when ramming gerbs etc. it is advisable to begin with one charge of starting fire especially where the composition contains steel, as they not only sometimes miss fire but there is also the likelihood of striking fire by ramming steel filings against a metal nipple.

GERBS.

These are used for all set pieces where brilliant effects or jets of fire are desired. They should be about ¾" diameter and 9" long. When steel filings are used the steel should be first protected q. v. as the saltpeter corrodes the filings which affects their brilliancy. They are

Fig. 48

rammed like rockets but on a short nipple without central spindle. Use the following formula.

STARTING FIRE FOR GERBS

Meal powder	4
Saltpeter	2
Sulphur	1
Charcoal	1

GERBS.

Meal powder	6	4
Saltpeter	2	
Sulphur	1	
Charcoal	1	1
Steel filings	1	2

FOUNTAINS.

Meal powder	5
Granulated saltpeter	3
Sulphur	1
Coarse charcoal	1
FF rifle powder	¾

FLOWER POTS.

Saltpeter	10
Sulphur	6
Lampblack	3
FFF rifle powder	6

CASCADES.

	1½" case	2" case
Granulated saltpeter	18	16
Mixed charcoal	4	4
Sulphur	3	3
Iron borings	6	7

WASP LIGHT.

Fig. 49

This contrivance a very effective and safe method of destroying the nest of wasps, hornets etc. The sketch illustrates the method of using same and the following composition is satisfactory.

Saltpeter	9
Sulphur	1½
Charcoal	5

Where it is not practical to attach the light as shown, a long pole may be used. Tied to the end of a fishing rod and brought in contact with a nest it will destroy it without danger to the operator as the burning composition completely demoralizes the insects who make no effort to sting.

The case should be about ¾" diameter and 5" long rammed on a nipple like a gerb. (Fig. 49).

REVOLVING PIECES.

Triangles.

These are made in two forms, (Fig. 50) (a) consisting of a small six sided block with concaved groves on three of its edges into which small choked cases are fastened, either by glue, wire or nailed; and (b) consisting of a triangular block on each side of which a serpent is fastened. The serpents must be rammed full and primed

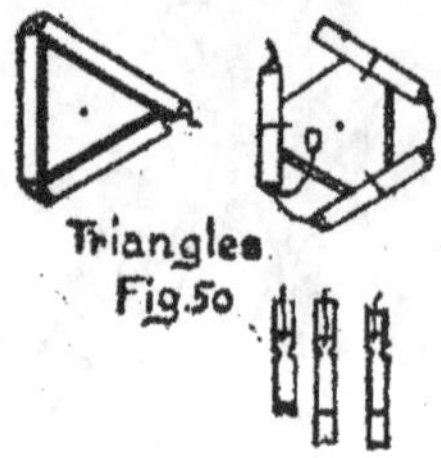

Triangles.
Fig. 50

at both ends, except the last one. A piece of paper is pasted over the joints where the two ends meet, of second and third cases and first one is matched with a small piece of match for lighting. The blocks have a hole through their center for the nail on which they revolve.

In making the larger triangles (a) take 3 small choked cases ½" diameter. Ram two of them with triangle composition to within ¾" of the end; then stop ends with ½" of same composition moistened with dextrine water and ram tight with solid rammer. The third case should be closed with clay. Now cut papering 2" longer than the case and cover in the regular way. Into the choked end after priming, twist a piece of match 1½" long, except the first one where a shorter piece will suffice. Fasten them to block as described above, first the one with clayed end, then one with both ends open and finally the one with short match. Insert match of third case into nosing of second one and match of second case into the first and secure the joints with pasted tissue paper.

VERTICAL WHEELS.

Made by fastening 4 to 8 driving cases to a wooden wheel made for this purpose. The cases are usually ½" to ¾" in inside diameter, either choked or rammed on a nipple with clay. They are papered and matched the same as for triangles except that the connecting matches should be papered as the distance between cases is greater than in the triangles. A little gum on the side of case

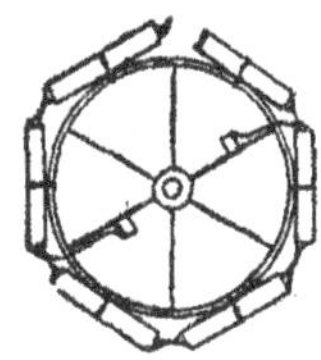

Vertical Wheel.
Fig. 51.

where it touches the rim of wheel will hold it more securely than wire alone. The wood wheels for these may be obtained in North Weare, N. H.

SAXONS.

Saxon Fig. 52.

Ram two cases ½" to ¾" inside diameter with a strong composition, closing both ends with clay and gluing them to a block as shown. Holes are bored ¼" diameter and just through case, as near to clayed ends as possible and at right angles with the nail hole in center of block on which saxon will revolve. These holes must, of course be on oposite sides. A piece of match is fitted into one of these holes and secured with pasted paper while another hole is bored into bottom of case but on side opposite to that of first hole. From this hole a piece of match is led to hole in second case; fastened with a tack and well secured with pasted strip. (Fig. 52)

Colored pots are attached to wheels and saxons greatly enhancing their beauty, by ramming light cases ¾" diameter and 2" long, with torch composition. They are fastened to the piece as shown and usually matched to the second case. Also, on larger vertical wheels the composition of the various drivers is varied so as to increase their effect as burning proceeds. The first case is charged with plain driving composition; the second with steel filings added; the third with granite stars etc.

Triangle Composition

Saltpeter	18	12
Sulphur	2	8
Mixed charcoal	5	5
Rifle powder FFF	12	12

Wheel Cases (Drivers)

Meal powder	8	3
Saltpeter	3	2
Sulphur	1	1
Mixed charcoal	1	1
F rifle powder	1	
Lampblack	½	
Steel filings ad. lib.		

Saxons

Meal powder	4
Sulphur	2
Saltpeter	2
Mixed charcoal	1

STARS.

This subject covers probably the most comprehensive division of the art of fireworks making. Besides the endless variety of colors, effects etc. we have the cut star, box star, pumped star, candle star etc. Nearly all stars are made by dampening the composition with water (if composition contains dextrine) or alcohol (if it contains shellac) and pressing the caked mass into little cubes, cylinders etc. by the various devices to be described.

CUT STARS.

These are the simplest form of stars in use. Secure some oak wood strips 1" wide and ⅜" thick, dressed, and from these make a frame about 12" wide, 18" long and $\frac{3}{8}$" high inside measurements, when lying down. The corners should be secured by halved joints, glued and fastened with small wire nails, clinched. Also provide a rolling pin about 2" diameter and 15" long. Now take any one of the formulas given for cut or pumped stars and moisten it rather more than for use with pump. The most convienent way to moisten any composition is to have a large dish pan or small wooden tub into which the composition is put while water is added little by little, working it in by rubbing the dampened portions between the hands until it is evenly moistened and a handful, firmly squeezed retains its shape.

Lay a piece of stiff cardboard on a marble slab, dust it with dry composition and lay on it the above mentioned wood frame. Fill frame heapingly with the dampened composition and press it down firmly with the rolling pin, leveling it off with a sliding motion so it is flush with the top of frame. Now, with a ruler and a table knife score the composition in each direction at a distance of $\frac{3}{8}$" apart, so as to cut it into cubes. This is facilitated if the frame has been previously marked at $\frac{3}{8}$" intervals. Make a cut around the inside edge of the frame so as to loosen the stars and carefully remove it. The batch may now be placed in the sun to dry. When throroughly dry the cubes may be broken apart for use.

On account of the ease with which these stars ignite, owing to their sharp corners, they are particularly adapted to rockets, small shells etc. where smooth stars are liable to miss fire. If larger sized stars are desired a frame of $\frac{1}{2}$" material or thicker may be used.

JAPANESE STARS.

This beautiful effect is made somewhat similar to the above. The great difference between the bulky lampblack and the compact potash makes it quite difficult to mix them thoroughly and this is particularly necessary to, obtain good results. Furthermore it is hard to get lampblack to take up water. It is therefore necessary first to moisten it with alcohol when it will take the water more readily. The method which I have followed with best success is as follows;

	#1	#2
Lampblack	12 ozs.	6 ozs.
Potassium chlorate	8 "	4 "
Saltpeter	1 "	
Water	18 "	9 "
Alcohol	4 "	2 "
Dextrine	1 "	
Gum arabic		½ "

Mix the dextrine and saltpeter (formula 1) well together and add sufficient water to make a gummy liquid. Boil the balance of the water and add the chlorate of potassium to it. Put the lampblack in a large pan and pour the alcohol over it working it in as well as possible. Now add the chlorate of potassium disolved in the hot water and stir with a stick until cool enough for the hands. Lastly add the dextrine and saltpeter. Remember that you cannot mix it too well and the effect will be in proportion to the evenness with which this has been done.

Take some pieces of light canvas or ticking about 18" square and put one or two handsfull of composition onto it; spread it about an inch thick in center of cloth, folding same over it, and place under a strong press of some kind. Fold up another cloth of composition in a similar manner and place on top of first. Repeat until 4 or 5 cloths are under press and screw up as tightly as possible and until surplus water runs out freely. Open press, remove cakes from cloths, dry for about two weeks and break into pieces about ½" square. It is important that the lampblack is perfectly dry and free from oil to get the best results and it is sometimes necessary to pack a jar or crucible with it and heat in a bright fire until all volatile impurities are expelled. You will then have one of the most beautiful effects of the entire fireworks art.

In recipe #2 the potass and lampblack are sifted together several times; add alcohol; then water in which gum has been disolved and proceed as in recipe #1.

BOX STARS.

Box Star. fig. 54.

Where the beﬆ and handsomeﬆ effects are required this form of ﬆar is undoubtedly the moﬆ adaptable to the purpose. Firﬆ, they burn much longer than others; second, they are less liable to go blind and furthermore they will ﬆand more blowing from a shell than any other form of ﬆar. (Fig. 54).

Make some light cases of about four thicknesses of ﬆout manilla paper 6" to 12" long on a ½" former. Cut with a scissors into ¾" lengths. Cut some thin match into lengths of an inch or a little over. Pass a piece of match through one of the little pieces of case or "pill box", bend the ends slightly around the edges as shown in illuﬆration and dip it into a pan of composition previously dampened as described before. Then with the firﬆ and second fingers of the right hand press the composition into it as firmly as possible until it will hold no more. Dry in the sun for two or three days.

PUMPED STARS.

Star pump.
Fig. 53.

These are used more than any other form of ﬆar on account of their regularity and the ease and speed with which they can be made, being even more quickly made than cut ﬆars where the proper appliances are at hand. Where only a few are required, a hand pump (Fig. 53) will do very good work. All that is necessary is to draw up the plunger, press the pump into damp composition until filled and by pressing the plunger while holding the tube, a ﬆar is ejected. When they are required in large quantities, however, star plates are necessary. With these 200 or 300 stars are made almost as quickly as one by hand pump. A good idea may be obtained by reference to (Fig. 55). The standard sizes of stars are about as follows:

#1	#2	#3	#4
¼" diam.	5/16" diam.	⅜" diam.	7/16" diam.
⅜" long	7/16" long	½" long	9/16" long

Consequently the plate for making a #1 star must be ⅜" thick and have holes ¼" diameter (a). The others, in same proportion. The plungers on plunger plate (e) must be somewhat smaller in diameter and slightly longer

Star plate assembly Fig.55.

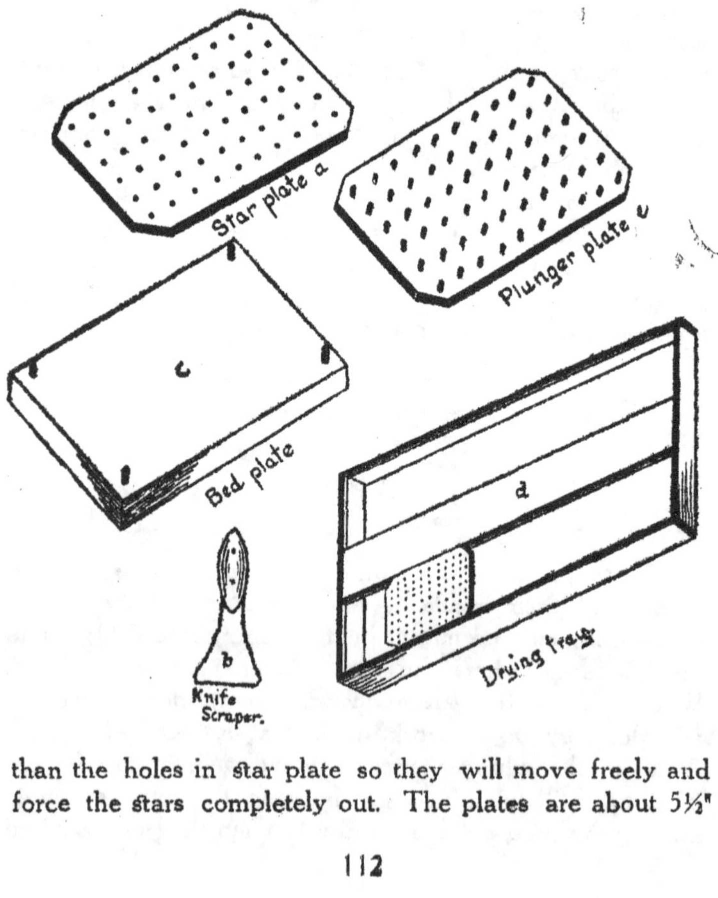

than the holes in star plate so they will move freely and force the stars completely out. The plates are about 5½" x 7½" square. Trays for holding the stars while drying (d) should have the bottoms made of brass wire netting so as to permit free circulation of air through the stars, enabling them to dry in a few hours. The center strip as well as the sides of the tray on top should be rabbeted so as to hold the star plate while stars are being pumped.

In order to make stars with a star plate, moisten a batch of composition in a dish pan with water as heretofore described, and empty same on a rather high work table previously covered with a square yard of rubber cloth. Press the plate (a) into same until the composition comes up through the holes. Then with the scraper knife (b) work more composition down into the holes until they appear full. Scrape off all surplus composition and remove to the iron bed plate (c) putting down the side previously up and press more composition in with the scraper. When all the holes are well filled scrape off surplus thoroughly, place in rabbet of tray and with plunger plate pump out the stars. Care must be taken to have right side of the plate up when pumping or plungers will not fit holes. If the plate begins to work badly on account of the composition drying on plungers same must be washed before using again. The proper dampness for composition can only be ascertained by practice. If too dry stars will crumble. If too wet they will not ignite freely. The holes in the star plate as well as the plungers may be much closer together than shown in cut.

Formulas for Cut, Pumped or Candle Stars.

WHITE STARS.

Saltpeter	50	54
Sulphur	15	15
Red arsenic	15	9
Dextrine	3	3
Black antimony		15
Red lead		6
Shellac		1

RED STARS.

Chlorate potassium	6	24
Shellac or red gum	1	3
Fine charcoal	2	4
Carbonate strontia		4
Nitrate strontia	6	
Dextrine	½	1½

BLUE STARS.

Chlorate potassium	24
Paris green	9
Nitrate Baryta	8
Shellac	5
Dextrine	1½

GREEN STARS.

Chlorate potassium	6
Nitrate Baryta	6
Fine charcoal	2
Shellac or KD gum	1
Dextrine	½
Calomel ad lib.	½

YELLOW STARS.

Chlorate potassium	16	16
Shellac or red gum	3	3
Fine charcoal	4	1
Nitrate Baryta	6	
Oxalate soda	1	7
Dextrine	1½	1

EXHIBITION PUMPED STARS.

GREEN (not for shells)

Barium chlorate	12
Potassium chlorate	8
Calomel	6
Shellac	2
Picric acid	2
Lampblack	1½
Dextrine	½

RED

(for hand pump; not suitable for Shells).

Nitrate strontia	8	
Chlorate potassium	4	10
Picric acid	1½	1½
Shellac	1½	¾
Fine charcoal	1	1
Dextrine	½	¾
Strontium carbonate		3

Exhibition Blue Stars, pumped.

Potassium chlorate	48	18	16
Calomel	18	6	12
Black oxide copper	6		
Asphaltum	6		
Dextrine	1½	1	
Paris green		4	
Stearine		2	
Copper ammonium chloride			4
Lactose			6
			*

* moisten with shellac solution

BOX STARS.

Red

Strontium nitrate	3
Potassium chlorate	3
Shellac	1
Dextrine	¼

Green

Barium nitrate	3	
Potassium chlorate	4	
Shellac	1	1
Dextrine	¼	
Barium chlorate		9

Blue

Paris green			25
Potassium chlorate	10		50
Potassium perchlorate		24	
Copper Sulphate	3		
Copper ammonium chloride		6	
Shellac	2		
Stearine		2	8
Asphaltum		1	
Dextrine	½		5
Calomel	2		

In first recipe, mix thoroughly the copper sulphate, shellac, calomel and dextrine; then add chlorate potash, previously sifted alone. This star is only suitable where it is to be used within a few weeks. In second formula rub up the stearine with the copper ammonium chloride in a mortar before adding other ingredients.

Pink

Potassium perchlorate	16
Plaster paris	4
Shellac	3

Yellow

Potassium chlorate	4
Sodium oxalate	2
Shellac	1
Dextrine	¼

White

Saltpeter	7
Sulphur	2
Powdered metal antimony	1½
Dextrine	¼

Purple

Potassium chlorate	18
Blk. oxid copper	1
Calomel	6
Strontium nitrate	1
Dextrine	½

LAMPBLACK STAR.

Meal powder	7
Lampblack	3
Black antimony	1
Dextrine	¼

Moisten with water; press into cakes, dry for one week and break into pieces about ¾" square.

SILVER SHOWER.

Saltpeter	50	18
Sulphur	15	6
Red arsenic	15	
Charcoal	10	¾
Dextrine	3	
Black antimony		6
Lampblack		1

Moisten with water.

GOLDEN STREAMERS.

Saltpeter	8	8	25
Oxalate sodium	4		
Sulphur	2		
Charcoal	½	4	7
Dextrine	½		1
Shellac			1
Lampblack		3	
Black antimony		1	

STEEL STARS.

Saltpeter	8
*Steel filings	2
Meal powder	1
Charcoal	1
Dextrine	¼

(*treated with parafine)

Moisten with water.

YELLOW TWINKLERS.

Potassium chlorate	8
Lampblack	12
Stearine	1½
Saltpeter	1

Moisten with alcohol and shellac. Pump with hand pump.

ELECTRIC SPREADER STARS.

The effect of these stars is quite suprising. A small pellet, no larger than a pea will spatter over an area of 15 feet when lighted. To make good electric spreader stars requires considerable care and judgement as dampening too much or too little greatly reduces their effectiveness.

Zinc dust	36
Potassium chlorate	7½
Granulated coal	6
Potassium bichromate	6
Dextrine	1

Mix thoroughly all but the charcoal and dampen until quite wet. Then add coal and mix again and pump with hand pump. Coal must be all coarse from which the fine has been sifted.

GRANITE STARS.

Saltpeter	14
Zinc dust	40
Fine charcoal	7
Sulphur	2½
Dextrine	1

For shells and rockets, this makes a very good substitute for electric spreader stars while being cheaper and safer to handle. It is moistened until quite wet, pressed into cakes ⅜" thick, cut into squares ⅜" each way, thoroughly dried and broken apart.

GOLD & SILVER RAIN. (cut stars)

Meal powder	16		4
Saltpeter	10	1	1
Sulphur	10	1	
Fine charcoal	4	1	2
Lampblack	2		
Red arsenic	1		
Shellac	1		
Dextrine	1		
Lead nitrate		3	

Moisten with water and cut into squares ¼" each way.

ALUMINUM STARS. (box stars only)

Potassium chlorate		8
Potassium perchlorate	8	
Aluminum powd. medium	4	4
Shellac		1
Lycopodium	1	

Moisten with shellac solution and form into box stars ¾" long, ¾" diameter.

MAGNESIUM STARS.

Saltpeter	5
Magnesium powder	2

Moisten with linseed oil. Owing to its high cost and unstability magnesium has been almost entirely replaced by aluminum.

COMETS.

These are large stars about 1½" diameter fired from small mortars of paper. In their simplest form they are just large pumped stars. If the gun is 10" long a piece of quickmatch 16" long is bared at one end about 1" and at the other 5". Lay it along side the comet star so the 1" bared end can be bent over the bottom. Then paste a strip of paper 4" wide and 10" long and roll this around the star over the match with the same amount projecting on each side. When dry gather the upper

Comet

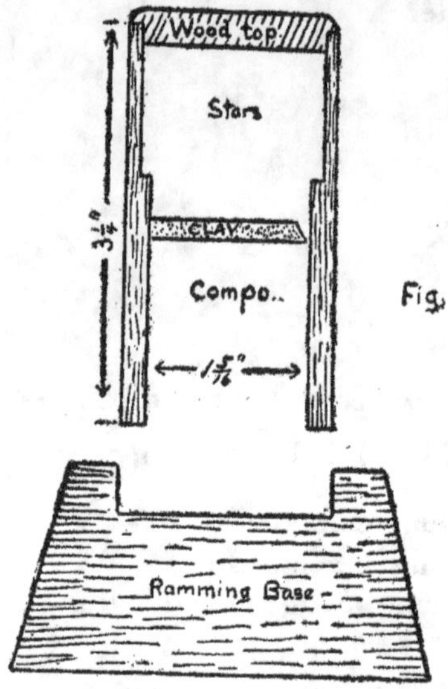

Fig. 56.

extension around the match with a string and into the lower projection or bag put a half teaspoonful of coarse grain powder and secure with a string also. Now drop this in the gun and it is ready for use. Handsome effects are obtained by making half of the star of red star composition and the other half of streamer composition.

A more ambitious form of comet is illustrated in Fig. 56. This is rammed into a case as shown, while the upper half, separated from the lower portion by a diaphragm of clay with small connecting orifice, is filled with small stars and blowing charge. At the end of its flight the stars are discharged with fine effect.

AEROLITES.

These are made by placing a comet star at the bottom of a short gun with blowing charge but no match. Over the star is placed 2" of candle composition and over this 1" of bengal fire.

COMET STAR COMPOSITION.

Saltpeter	6	
Meal powder	6	3
Sulphur	1	
Fine charcoal	3	1
Powdered antimony	3	1
Lampblack		2

LANCE WORK.

This is a division of pyrotechny which consists of reproducing with colored lights various designs, portraits, lettering etc. after the design is sketched on the floor as described under "Designing".

Fig. 57

A number of frames are made, 5' wide and 10' long, of light lumber ½" x 2" for the outside strips and ½" x 1" for the center ones, spaced 1 foot apart each way with a brace in the two corners as shown in (Fig. 57). These are laid over the design on the floor and secured so they do not shift until completed and the picture etc. transfered to the frames with bamboo for the curves and light strips of wood for the straight lines a and b Fig. 58.

When this has been completed frames should be numbered, begining at left hand upper corner of first frame and numbering each consecutively to assist in getting them in their proper places when erecting to be burned.

The frames are now ready for attaching the lances. This is done by driving a 1¼" wire nail to a depth of

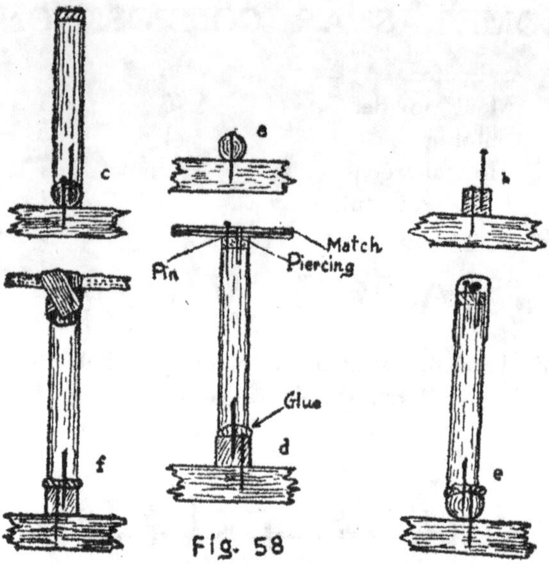

Fig. 58

half an inch, (b) at intervals of 2½" in curves and 3 to 4" on straight lines all over the design. Be sure to see that there is always a nail at every point where two lines intersect. Now with a nipper cut off the heads of the nails, holding the nipper at an angle with the nail so that the place where the head has been cut off will have a sharp point instead of being cut off square across.

The frames are now ready for the lances. When it has been decided what colors are to be used for the various parts of the design, take a handful of lances of the desired color and dip their bottoms into glue to a depth of about ⅛" and press one onto each of the nails until they are attached firmly to the cane or sticks forming design, (c) (Fig. 58).

When glue has hardened frame is ready for matching. Take a length of quickmatch and, begining at the upper end of frame pin it from one lance to another until entire frame is covered, following the outline of design as much as possible, (f) (Fig. 58). When the end of a length of match is reached splice another to it by baring about 3" of the new length and slipping this bare end into the pipe of the preceeding length; securing by tying and pasting joint.

Leave an end of match about 2' long projecting from the lower right hand corner of each frame so it can be connected to the one next to it when erecting. Also on

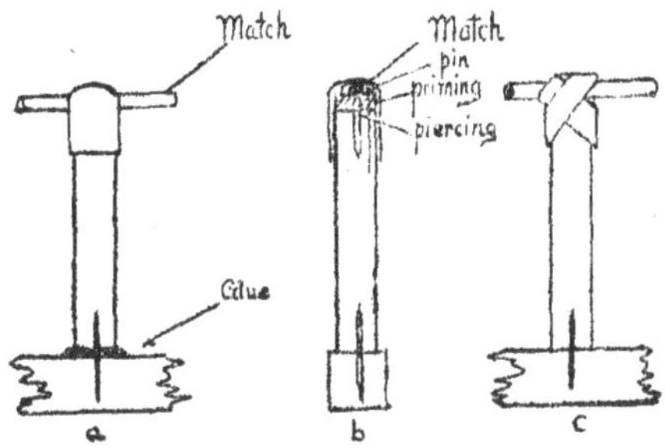

Fig. 59

one of the bottom frames leave a leader 10 to 20 feet long, of match, to light piece by when it has been erected. Now with a three cornered awl make a hole ½" deep through the match pipe and into the priming of each of the lances on the frame, (d) (Fig. 58). Then take strips of tissue paper ½" wide and 3" long; paste a number of them onto a light board and working along from lance to lance secure the match to top as shown at Fig. 58. Sometimes where it is desired to rush a job to be burned same day at point where it is made, the lances are secured by simply bending a pasted strip an inch wide over top of lance as shown at a, b, Fig. 59.

The completed frames may now be crated into lots of four, with ½" x 4" strips arranged to hold them apart, for convienence in transportation.

LANCES.

These are small paper tubes from ¼" to ⅜" diameter 2" to 3½" long filled with composition burning different colors with a duration of one minute and used for producing the different designs used in fireworks exhibitions, such as portraits, mottoes etc. The cases are rolled and rammed with funnel and rod, as previously described.

Some lance compositions are so light as to be difficult to ram. These should be slightly dampened first. Blue lances made with paris green and white ones using realgar are frequently used without priming as they ignite very easily.

A good lance should burn clear for one minute, without flaring or clogging up. All colors should burn of about the same duration. If a lance burns to one side it is often because the composition is not well mixed or because there is more paper on one side than on the other. They should have about three turns of paper all around.

LANCE COMPOSITIONS.

Red Lances.

Potassium chlorate	16	16
Strontium nitrate	3	
Strontium carbonate		3
Shellac	3	2
Lampblack	1/8	1/4

Green Lances.

Potassium chlorate	7	16	
Barium nitrate	7	4	4
Barium chlorate			5
Shellac	2	4	1
Calomel		3	2
Lampblack		1/8	
Picric acid		1	

White Lances.

Saltpeter	9	14	5	8
Sulphur	1	4	2	2
Antimony sulphide	2			
Antimony metalic		3	1	
Meal powder			1	
Red arsenic				1

Blue Lances.

Potassium chlorate	20	16	12	
Potassium perchlorate				24
Paris green		5		
*Copper sulphate	6			
Copper ammonium sulphate			3	
Copper ammonium chloride				6
Shellac	4		1	
Stearine		1½	½	2
Calomel	4	3	3	
Dextrine	1			
Asphaltum				1

*See directions under box stars for using this.

Yellow Lances.

Potassium chlorate	16	4	4
Sodium exalate	2	2	2
Shellac	3	1	1
Charcoal	½		
Barium nitrate			1

For amber and purple lances the recipes given under TORCHES may be used to advantage.

BOMBSHELLS.

These represent the highest developement of the pyrotechnical art and require great patience and skill for their successful production. The most wonderful effects are produced by the Japanese while the finest color effects are made by the Europeans and Americans. Shells are made in several forms though round is the most popular. Cylindrical or canister shells however contain more stars etc. and in the more complicated effects it is sometimes necessary to attach a canister to the round shell to contain the parachute etc. (Fig. 64).

Shells are all fired from mortars, the smaller ones made of paper, up to about 3" diameter and the larger ones of wood, copper and iron. The smallest shells with which we have to deal are the

FLORAL SHELLS.

	#1	#2	#3
Diameter	2-5/16"	2-7/16"	3-3/16"
Height	9"	11"	13"

The shells are made of hollow wooden balls which can be turned out by any wood turner. They are made in halves, usually with a rabbet to insure a close fit. Through one half, drill a hole just the size to snugly fit a piece of ordinary blasting fuse, 1¼" long. Glue the fuse on the inside as well as on the outside of the shell case. Now fill each half with stars to which add a teaspoonful of shell blowing powder, glue the edges of each half, clap them together and when dry paste a strip of paper around where the two halves join. Prime the ends of the fuse which should project through the shell about ¼", bend a piece of naked match, about 8"

Fig. 60

Floral Bombshell.

long, around the shell so that the middle of it passes over the fuse, tacking the ends to other half of shell, so that they will stick over about 2". Set it now on the floral shell bottom as shown at Fig. 60 into which has been previously put an even teaspoonful (for the #1, larger sizes in proportion) of F grain powder, and secure with a strip of pasted paper. When dry, slip over the paper mortar, having previously well glued the bottom. Measure distance from top of mortar to top of shell inside and mark same on outside. Punch hole through same at this point; fit on a top and secure. Now take a roman candle a little longer than the mortar; punch a hole in its side near the bottom star; fit in a short piece of piped match bare the other end; slip into hole in mortar, fasten candle to side with wire and floral shell is completed.

METEORIC SHELLS.

These are made somewhat differently; ¼ lb. to ½ lb. tin cans being substituted for wood shells. They are filled with stars, colored and Japanese and bursting powder in the same manner as described above. A hole is punched through lid into which fuse is glued. A strip of paper about 3" wider than the length of can and long enough to roll around it six times is pasted all over, the filled

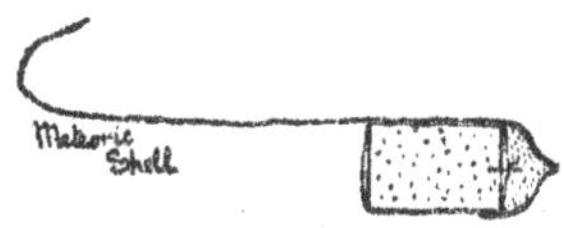

Fig. 61

can placed on it and rolled up something like a case is rolled. The 1½" projecting over each end is now carefully pressed around the fuse on one side and the can bottom at the other. The shell is allowed to dry for a week before using. The end of fuse is trimmed and primed; a piece of piped match bared at each end is laid against its side, extending 1" beyond fuse at bottom of shell. A nosing 3" wide which secures the match in place is attached to shell and this when dry serves to contain the driving charge of a level teaspoonful of grain powder, after which it is gathered together and tied with twine. A dab of glue on bottom of bag suffices to hold it in bottom of mortar and it is ready for use. (Fig. 61).

EXHIBITION BOMB SHELLS.

The principal sizes of shells used for this purpose are 4", 6" and 10" diameter. For round shells, after the cases have been made as described under "CASES" the upper halves are bored for the fuses. This may be done with a carpenters brace bit ¾", boring from the inside. Fit the two halves accurately together; bind with a strip of glued cloth and over this two or three layers of paper strips laid on longitudinally; each strip overlapping the one before it by about ¼". If each layer is made of a different length a better finish will be obtained.

When cases have been thoroughly dried fill them with the desired stars through the fuse hole. When they will hold no more add blowing charge. The fuse should now be accurately fitted by cutting around it with a knife ¾" from the top and peeling off a layer or two of paper until it will just enter the hole which has been made in top of shell for it. Glue lower portion well and push into place until shoulder rest squarely against shell case.

Attach a cloth nosing to fuse; bare 1" at the end of a length of match and attach to bottom of shell by a #1 tack; lead match up to fuse bend at right angles to permit of entry into nosing; cut the piping at point of entry and secure with strong cord. The remaining match also serves for lowering shell into mortar, up to 6" sizes. Larger shells must have a heavy cord passed around them for this purpose.

The necessary driving charge having been placed in a paper cone this is attached to bottom of shell when same is completed.

Fig. 62.

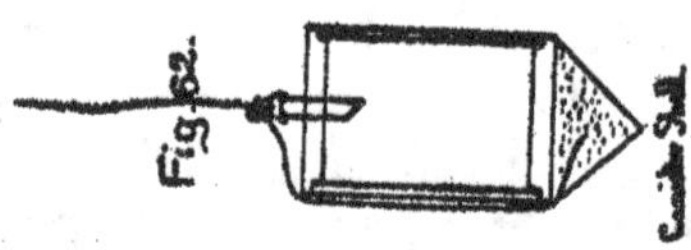

The making of canister shells is so similar that their construction can be readily understood from the sketches (Fig. 62). The heads and bottoms are made of ¾" wood.

Another method of matching shells is to start at the fuse by baring a half inch at the end of a match pipe and pushing this into nosing. Bend match at right angles ½" above nosing and pass under and entirely around shell, coming back again to nosing. Bend once more at right angles and insert bend alongside of where start was made first cutting through match pipe at point of insertion. Gather nosing closely around match and tie tightly as possible. This method gives a somewhat better support to shell when lowering it into mortar. Where the match crosses bottom of shell and enters driving charge be sure to cut piping away for about half an inch, (Fig. 63).

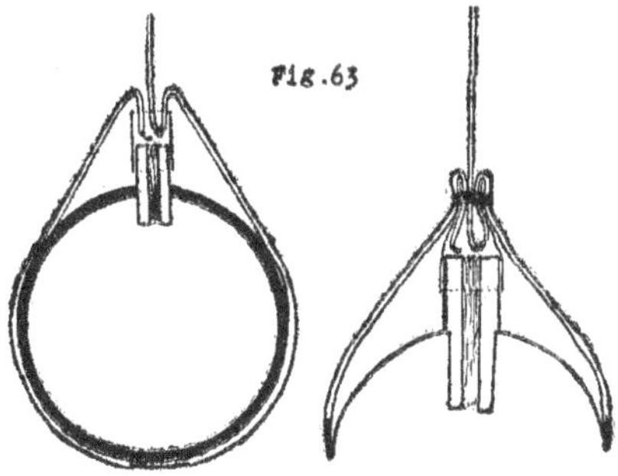

Two and three break shells are made by lightly fastening together the desired number of short canister shells with fuses not over ¼" long between them. The first shell has regular length fuse. The details can be better understood from drawings than from a description. See Fig. 65.

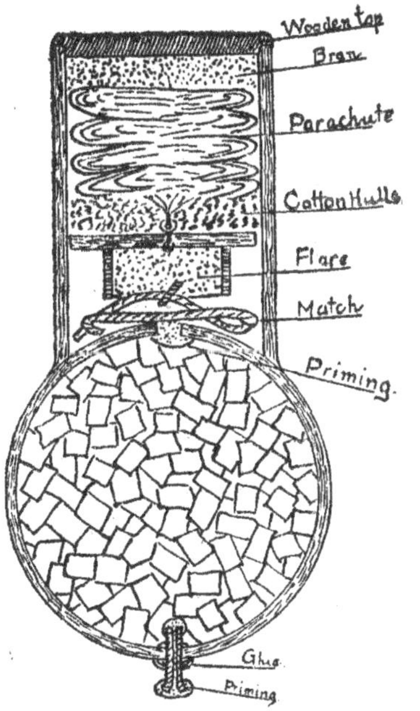

Parachute Shell. Fig. 64

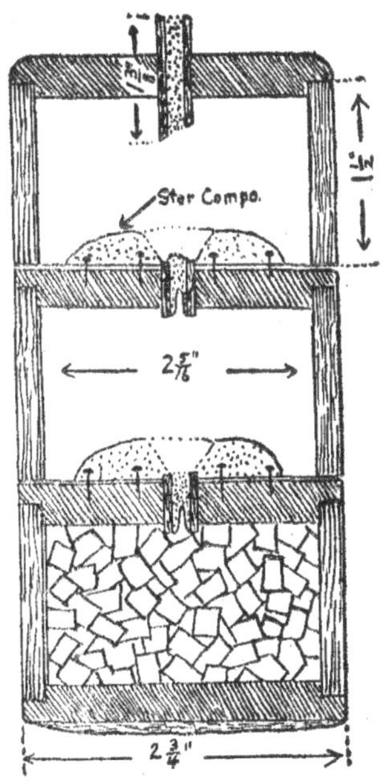

3 Break Shell. Fig. 65.

PARACHUTE BEARING SHELLS are also shown in detail, Fig. 64.

SHELL FUSES.

These are best made of hardware paper. Take a rod ⅜" diameter (for a 6" shell) and a sheet of paper 6" wide. Paste it with thick paste all over one side and at once roll it up as tightly as possible until it has an outside diameter of ⅞". The length of sheet required is dependent on the thickness of the paper. When a number of these cases are rolled they must be dried in the shade until they are as hard as wood and rattle when struck together.

Take a rammer 7/16" diameter a light mallet and some fuse composition made as follows:

Meal powder	4
Saltpeter	2
Sulphur	1

Sift and mix three times. Rest a fuse case on a firm block, scoop in a little composition and tap it about ten light blows. Add more composition and ram again, repeating until fuse is filled. The composition in fuse must be as hard as possible when finished, otherwise it will blow through when used in a shell. The fuse may now be cut into the required lengths with a fine toothed hack saw. (Fig. 66).

Length of fuses.	Inside diameter of fuses.
4" shell 1¼"	4" shell 5/16"
6" shell 1⅜"	6" shell ⅜"
10" shell 1½"	10" shell 7/16"

In some cases a hole is drilled into the composition of the fuse on the end in the shell, ¼" deep, so the fire from same is thrown into the shell with more force. In this case allowance must be made when cutting length of fuse.

Shell Blowing Powder may be made of:

Grain powder	1
Meal powder	1
Saltpeter	3
Charcoal	1½
Sulphur	1

The bursting and driving charge for shells is as follows:

Bursting Charge.		Driving Charge.	
4" shell	1¾ oz.	4" shell	1½ oz.
6" shell	5 oz.	6" shell	3½ oz.
10" shell	16 oz.	10" shell	14 oz.

The driving charge should be coarse grain powder; cannon powder is best. An endless variety can be produced with shells some of which follow while the ingenuity of the pyrotechnist will suggest others as he progresses.

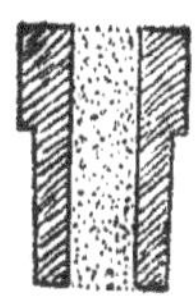

Fig. 66

Solid color shells
Varigated shells
Gold rain shells
Japanese or willow tree shells
Streamer shells
Aluminum shells
*Conch shells
Chain shells
Repeating shells
Maroon or Salute shells
Day shells.

*The Conch Shell consists of a 10" diameter shell packed with three ball roman candles made specially for this purpose. The cases of the candles are made of very strong paper so that they can be thin and no empty portion is left at top and bottom. In addition to the little roman candles, colored stars are added. This makes a very effective shell.

SHELL CONES.

(for holding driving charge)

These are made by cutting out a round piece of good Kraft paper six to twelve inches diameter according to size of shell for which cone is intended. With a scissors make a cut from the edge to the center and twist it around so as to make a bag or cone of two thicknesses, pasting the edges where they meet. Put the driving charge into this and with a little paste attach it to the bottom of shell having previously cut the match piping where it crosses the bottom, so that fire will strike driving charge when shell is lighted.

JAPANESE BOMBSHELLS.

Day and Night

The Japanese have developed this form of pyrotechny to an almost incredible degree of beauty and originality. Some of their shells are marvels of patience, ingenuity and skill.

DAY SHELLS consist of two kinds. First, those containing large figures of birds, animals etc. made of light tissue paper sewed together like a bag and open at the bottom with a row of small weights around the rim of the bottom. The figure is folded into a small compact pile and packed into a cylindrical shell case, somewhat as parachutes are placed in rockets. When they are fired to a height of about 1000 feet the figure is expelled with a light charge and as it falls, the weights cause the bottom to unfold and the inrushing air inflates it. One of their day shells contains about a dozen paper parasols which, of course. are folded when inside the shell case but which by an ingenious construction, open as soon as shell breaks and they float to the ground much as parachutes do. The arrangement is as follows:

The second variety of day shells consists of colosal spiders made of smoke and varicolored clouds which must be seen to be appreciated. They are made by filling

Fig. 66a.

Fig. 66b.

a round shell with smoke stars, on top of which is set a canister containing a number of 1½" diameter colored smoke shells and a parachute from which hangs a "smoke dragon".

Fig. 66c.

The night shells embrace some 50 or 75 different effects. Up to 40 years ago colors in night shells seem to have been unknown to the Japanese and all their devices consisted of endless varieties of tailed stars, gold and silver rain, willow trees and bright work, each one, more entrancing than the other. Among some of their very unusual effects is a night shell which upon reaching the height of its flight throws out five red paper lanterns with a light burning inside each one. The lanterns when inflated are about two feet in diameter and four feet high. When folded inside the shell they occupy a space about five inches diameter and nine inches long. Another of their original shells breaks with their very round effect but only one half of the circle is filled out with stars while the periphery only, of the other half is outlined with stars as shown below. (Fig. 66d). To secure this effect a wing or tail is attached to the shell which holds it in the required position, relative to the observers at the moment of explosion.

Fig. 66d.

The arrangement of the contents of these shells is shown in the accompanying sketch, (Fig. 67). The inside of the round shell case is scored as shown at (a) and (b) by cutting half way through the paper with a knife in order to cause shell to burst evenly and throw the stars equally in all directions.

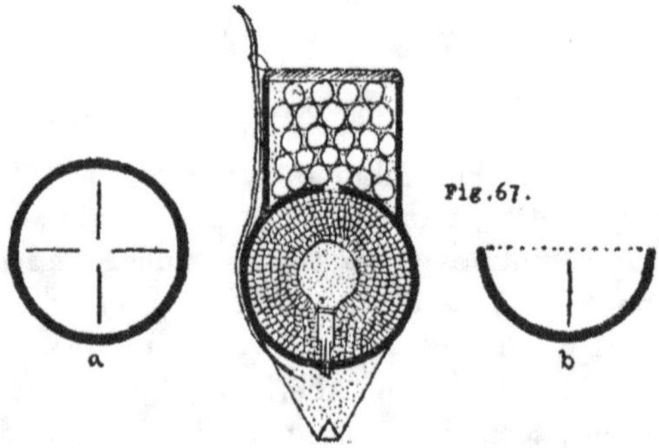

Fig. 67.

Smoke stars are pill boxes ⅝" diameter and ¾" long,

closed at one end and over the other end are bent two strips of thin metal (copper or tin) which are secured by a paper fastening so as to restrict the opening to about one third its original size. The stars are charged with the composition shown under "Smoke Stars", the end being well primed over the metal strips. They are then arranged in the round part of shell as shown in Fig. 67. The little smoke shells and parachute are placed in upper or canister portion.

MORTARS.

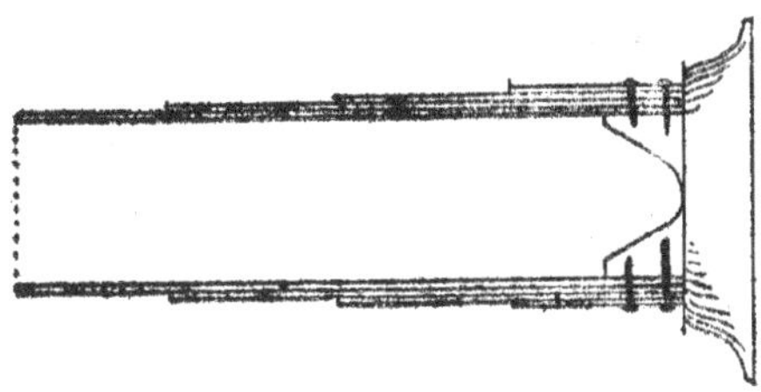

MORTAR
Fig. 68

Mortars for firing pyrotechnical bombshells are made in a variety of different ways. For shells up to 3" in diameter a mortar 12 to 15 inches high, made of a number of turns of good stout paper will serve for perhaps a hundred shots, especially if lined on the inside with a piece of tin or galvanized iron. If a bottom of oak or other hardwood is fitted to it and the barrel tightly wound with marlin, it will be perfectly safe, light and cheap.

For 4" diameter shells and upward mortars of copper tubes, shrunk, one over the other so that there are four thicknesses at the bottom, three for half the length two up to three-quarters of its barrel and one thickness for balance, with a ring at the top make an ideal mortar.

Wrought iron tubes wound with galvanized wire and fitted with cast iron bottoms securely fastened in by machine bolts or rivets make very servicable guns. The bottom should be conical on the inside to accomodate the powder bag of shell. (Fig. 68).

The Japanese used long wooden mortars reinforced with iron bands. These were soaked in water before being used, to swell and tighten them. On account of their length they would throw a shell to a great height with a small driving charge. They should always be buried for half their length in the ground when in use. With these mortars it was customary to pour the driving charge for shells, loosely into the mortar, drop the shell over it and fire by shaking a little dross from a port fire into the mortar. This method is dangerous and is not to be recommended.

Mortars with port holes on side of bottom, like the old style military mortars are sometimes used for day shells. The cartridge of shell is pierced with a priming rod and a piece of bare match inserted through port hole.

BALLOONS.

Secure some good tissue paper 20" x 30" (Foudrinier is best). Paste two sheets together on the 20" ends making a sheet 20" x 60". Split this lengthwise and get a sheet 10" x 60". Make twelve sheets of this size; lay one on top of another and double over the lot longitudinally so as to have pile 5" x 60". Now, with a scissors cut along the unfolded edge as shown in Fig. 69 (a), removing the part shown as shaded. The exact line to cut may be determined by practice until the most satisfactory shape is found when an extra sheet of heavy paper should be cut and reserved as a pattern. Or, the pattern can be made according to instructions given under "Balloon Designing".

Unfold sheets cut as above and lay one on the table before you. On top of this lay another but about ½" nearer to you thus leaving an uncovered edge of the under sheet exposed (b). Apply paste lightly to this edge and lap it over onto the upper sheet in this manner joining the two for their entire length. Make six pairs of sheets like this and then repeat the process with the double sheets until you have three sets of four sheets. Join these as before making the final closing joint

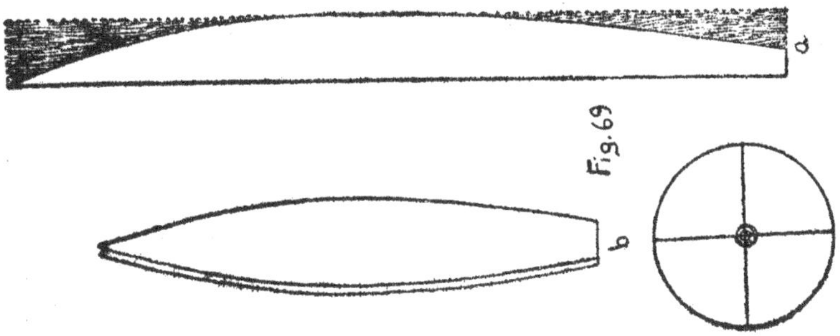

likewise. If the top of balloon where the joints meet is not well closed paste a small round piece of paper over all.

When balloon has dried make a ring of wire, bamboo or rattan for the bottom with cross wires to hold the inflator (c). For a balloon of the above size the ring should be about 15" diameter. In balloons 10 feet or more in height a wire basket is sometimes woven into the center of the ring so that an extra inflator may be added just before releasing balloon when ready to rise.

BALLOON INFLATORS.

These are made in several ways. One consists of a ball of cotton wool which is saturated with alcohol or kerosene oil when balloon is to be inflated. A more convienent inflator may be made by impregnating a ball of excelsior with parafine and fastening it on top of cross wires of balloon ring. This has the advantage of being cleaner and requires nothing further than lighting when balloon is to be raised.

DESIGNING BALLOONS

A balloon five feet high when deflated can be made from twelve pieces of tissue paper cut out of sheets 10" wide and 60" long. To get the proper shape for cutting these sections draw a plan of the desired shape of balloon when finished, somewhat as shown in Fig. 1. Then make a ground plan as shown in Fig. 2. Quarter the elevation plan by the two lines a1 and a2. The

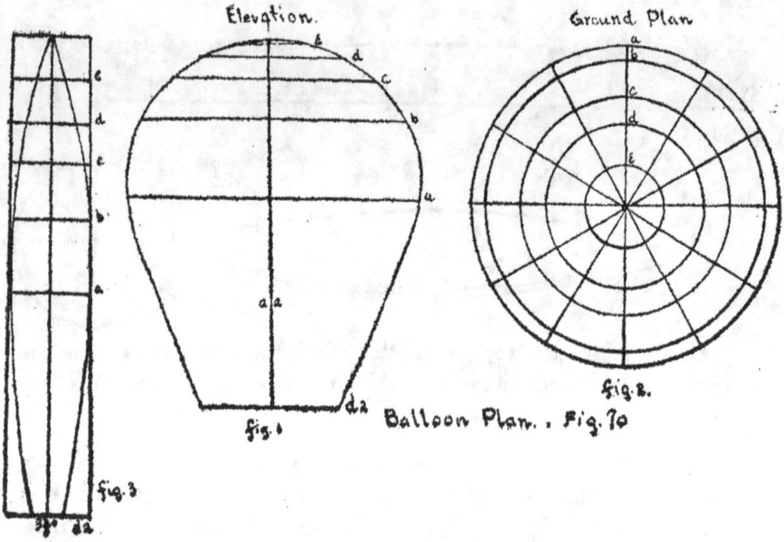

Balloon Plan. . Fig. 70

lines a1 represent the balloon at its widest point in both plans. Line b in ground plan is obtained by measuring the length of line b in Fig. 1 from central line a2 to the edge of balloon and then taking the same distance from the center of Fig. 2 and making a circle with a pair of compasses at this point. Lines c, d and e are obtained in the same manner.

Now, to make the pattern as shown in Fig. 3 draw a plan of one of the sheets from which the balloons is to be cut using same scale as in Figures 1 and 2. Divide it by a line through its center lengthwise and then locate lines a 1, b 1, c 1, d 1 and e 1 by measuring distance from bottom of balloon to each cross line on Fig. 1 along one edge from d2 to e. It now only remains to locate the points on Fig. 3 for getting proper shape of pattern. To do this take a pair of dividers and measure length on line b 1 from central perpendicular line in Fig. 2 to point where it intersects the next radial line to the right. Divide this distance equally to each side of central line of line bl in Fig. 3. Do the same with lines c, d and e. On a large plan this may be more accurately done with a flexible rule but when using dividers as above a slight allowance must be made for the curvature of the lines on Fig. 2. All that is necessary, now, is to draw an easy line from top to bottom of Fig. 3 as shown. The bottom of 5' balloon should be about 15" diameter. Dividing this by four will give approximately 3¾" for bottom of pattern.

Fireworks Attachments for Balloons.

These make a very pretty addition to a balloon ascension and may be designed in numerous ways as to ingenuity of the pyrotechnist will suggest. A typical attachment is shown in accompanying sketch (Fig. 71).

Fig. 71

The lower portion of a gerb is filled with red fire which burns until the balloon reaches the height of several hundred feet; then the gold rain effect of the gerb functions until the heading of stars, serpents etc. is discharged. A vertical wheel suspended from a wire and lighting when balloon is well up in the air makes a very interesting display.

CANNON CRACKERS.

In this item of pyrotechny the history of fireworks in general has been somewhat reversed. While many persons have lost limbs and life in the manufacture of roman candles, rockets etc. on a large scale, comparatively few serious accidents have occured to those using them. On the other hand, though cannon crackers are one of the safest articles in the business to make, they have caused, during their short career the loss of more hands, arms etc. to those firing them than all other kinds of pyrotechnics combined.

The reason is simple. The composition of crackers is only explosive when confined or after the cracker is finished and the explosion of a finished cracker will not ignite others, as is the case with candles etc. where a spark will fire thousands at once. But when crackers are used by the inexperienced it is difficult to determine whether the fuse is lighted or not which often causes the cracker to explode in the hand with disastrous results. Its bloody record has caused a number of states to legislate against its sale in sizes larger than three to four inch. Until a fuse is invented that will be consumed as it burns, this piece of fireworks will be dangerous to handle. The first available record of the manufacture of American Cannon Crackers on a commercial scale was about the year 1880 when Edmund S. Hunt of Weymouth, Mass. devised a very ingenious machine into which was fed empty cases from a hopper while the composition and fuse was inserted and the ends crimped at one operation. Previously only Chinese crackers were used but the increased loudness of the report and the reduced cost of making them soon caused the American article to supplant the imported one in the larger sizes.

With the advent of the Flash Cracker the Chinese have again invaded the American market to a large extent.

The cases for crackers are rolled similar to rocket cases except that paste is used only on the last turn at farthest end of sheet, the body of the case being rolled dry. By this means the cracker is more easily blown into small fragments and the danger of being struck by a large piece of hard case is avoided. The fuse used is the small red cotton untaped fuse made especially for this purpose though most any kind of blasting fuse may be used. A piece from 1½" to 3" long is sufficient according to the length of the cracker.

Various compositions are used, those containing Antimony giving the loudest report while those made with sulphur only, produce less noise. The cases should be filled about one third full of composition to obtain the best results and composition must be loose, not rammed. The addition of charcoal will increase the lightness of the composition and prevent its tendency to pack which lessens the report.

The ends of crackers are stopped in various ways. The best is by means of crimpers which pinch or mash the ends of the case around the fuse at one end and into a bunch or lump at the other. A dab of glue retains the ends in place. Another method is to close the fuse end with clay and the other with a cork. The low grade of corks used for this purpose can be bought for as cheaply as twelve cents per thousand in the small sizes which is less than the cost of plugs of any other sort.

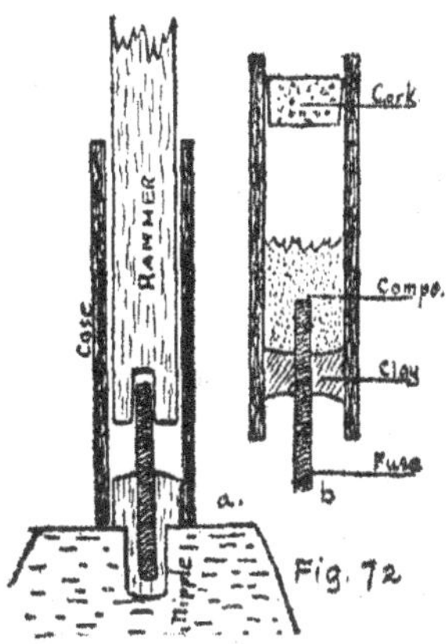

Fig. 72

To make crackers in this manner roll the cases as directed. Make a brass nipple as shown Fig. 72, of the diameter of the cracker desired and about as high from top to shoulder. Drill a hole through the center somewhat larger than the fuse so it will pass easily into it. Also provide a rammer about 6" longer than the cracker and drill a hole into the lower end, somewhat larger than the fuse and ream or countersink it a little. After setting the nipple in a block put a piece of fuse in it, slip a case on, put in enough slightly dampened clay to occupy a half inch when rammed and with a few blows of a mallet set it firmly. You had best now remove the case and with a sharp knife split it open without breaking the clay and see if everything has been operating correctly, that is, if the proper amount of clay has been used and if the fuse projects sufficiently on the inside and has not been mashed into the clay as sometimes happens if the rammer has not been correctly made on the end which rams the clay while the fuse enters hole in same. Unless this has been well done the cracker will fail to explode. When these matters have been properly adjusted proceed with ramming clay in another case and scoop in enough composition (any one from list to follow) to fill case about one third. Then take a cork that will fit pretty snuggly, dip the small end in liquid fish glue and force it in open end of cracker. It is now completed and may be removed from nipple. If too much composition is used the report is weakened; a full case will hardly explode at all.

When crackers are made on a large scale, a block of six dozen nipples is used, six wide and twelve lengthwise, same as for roman candles. The clay and composition can be dropped into lot simultaneously by the use of shifting boards as shown under candle machine. Some manufacturers prefer to use a long nipple and short rammer, reversing the manner of ramming as the cracker is rammed from the fuse end instead of from the cork end. By this means the danger of mashing the fuse into the clay is avoided as the nipple on the inside protects it. But only the clay can be filled into them all at once by this method, as, the composition being loaded from the other end, they must be removed from the spindles before this can be accomplished.

The following are the standard sizes.

Numbers		Length	Bore	No. in box	Boxes in case
1	Salutes	2"	5/16"	100	20
2	"	3"	5/16"	50	25
3	"	3½"	⅜"	15	100
4	Crackers	4"	7/16"	30	20
5	"	5"	½"	20	20
7	"	6½"	⅝"	10	20
9	"	8"	¾"	5	20
10	"	9½"	⅞"	3	20
12	"	10½"	1"	2	20
15	"	13"	1¼"		25 in case.

CANNON CRACKER COMPOSITION

Potass: chlorate	60	6	6
Washed sulphur	23	3	2
Sulphuret antimony	5		
Metalic antimony			1
Charcoal		1	
Saltpeter	12		

If unwashed sulphur is used the report will be considerably louder but the danger is greater. Of the above mixings, the last is about the safest that can be made The first gives the loudest report. Great care must be exercised in mixing the composition for cannon crackers. Each ingredient must be sifted separately and then mixed in a tub with the fingers, preferably gloved, being careful not to scratch the bottom of tub with the nails.

MAROONS.

The name is probably derived from the French word for chestnuts which burst when being roasted, and covers what consists of a case of heavy paper containing an explosive charge which when ignited produces a loud report. Aerial Maroons are simply the same article arranged to explode in mid-air when fired like a shell.

FLASH CRACKERS.

This interesting addition to pyrotechny is one of the results of the advent of aluminum. The following composition may be used both for flash crackers and maroon shells.

	#1	#2	#3
Potass: perchlorate		50	30
Potass: chlorate	6		
Washed sulphur	3	25	30
Pyro aluminum	1	25	40
Charcoal	1		

Mix thoroughly the sulphur, coal and aluminum; then add the chlorate potass; previously sifted by itself; mixing by rolling the ingredients back and forth on a piece of paper and avoiding friction of any kind.

Prepare a block c Fig. 73 by boring several holes as shown, 7/16" diameter and 1" deep. Also a nipple ⅜" diameter fitted into a handle b and some pieces of strong light paper 2½" square a. Take a piece of paper in one hand and with the nipple in the other, press the paper around it so as to form a little cup which is now inserted in the hole in block pressing down until the

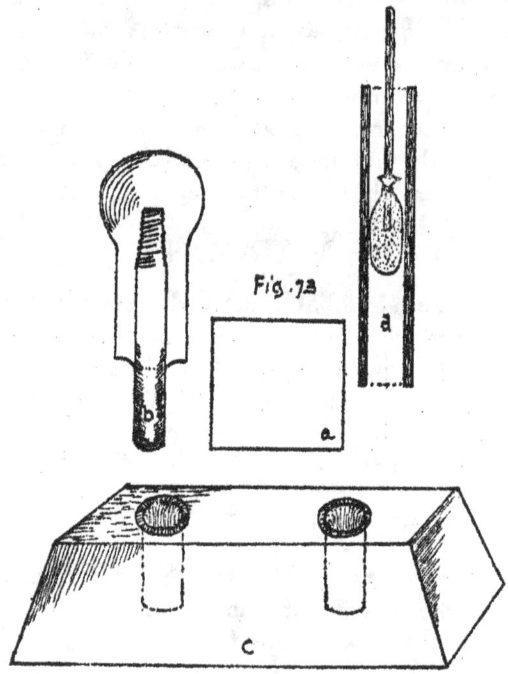

flange of the nipple spreads the upper edges of the paper. Remove the nipple and put into the paper cup formed, enough composition to half fill it. Insert a piece of match three inches long; draw the paper around the match and secure tightly with two half hitches of linen twine. Remove from block, smear a little gum on one side and push into a cracker case (d) ½" diameter and 3" long. It is not necessary to close the ends of flash

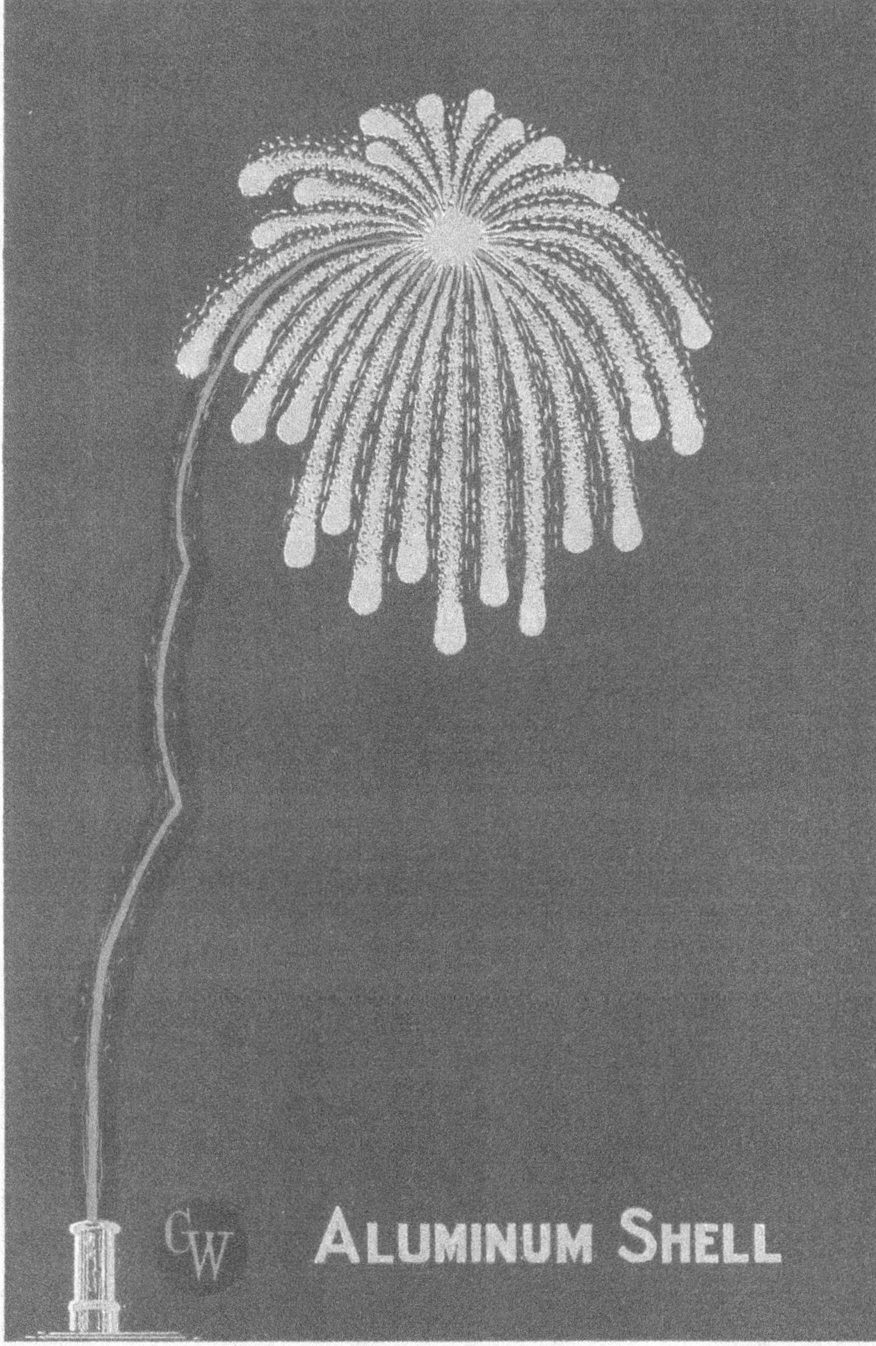
GW
ALUMINUM SHELL

crackers as the report is just as loud when they are left open.

Composition #2 and #3 are much the safest and should be used by any but those very familiar with this class of work but the ends of the cracker must be tightly closed to secure a loud report as directed for cannon crackers. Also the sulphur in these mixings should not be washed.

TORPEDOES.

By this name is understood the toy torpedoes used by children, which detonate when thrown on the ground. I believe that these were first made by the French under the name of "Pois fulminant" (mad peas), but the so called Jap. or cap torpedoes, which constitute far the largest part of those used today, as nearly as I can find out are an American invention.

SILVER TORPEDOES.

Fulminate of Silver.

The fulminate is prepared by taking 8 ozs. of C. P. Nitric acid (42%) and adding gradually, stirring constantly with a glass rod, 2 ozs. water. Into this put a silver dollar or 1 oz. of bar silver. Warm slightly until a brisk reaction takes place. When the silver is completely dissolved allow to cool for three minutes. Then add 16 ozs. pure alcohol. Add it all at once and quickly and be sure that the vessel containing the solution of silver is at least four times as large as the amount contained for a violent effervesence will take place. After it subsides add three ounces more alcohol. Let stand for a quarter to one half hour when a white crystaline precipitate will be formed on the bottom of vessel which is the fulminate and may be collected on a filter and dried in a shady place. A candy jar may be used for making the fulminate but a glass beaker is far preferable.

The utmost care must be excercised in handling the dry powder as the slightest concussion will explode the entire lot with terrific violence. A wooden spoon should be used for removing it from the filter and it should be handled as little as possible and in the smallest practicable quantities. Procure a round paper box from a drug store, one inch in diameter and three inches high. Make a small hole 1/16" in diameter through the cover and fill it about half full of dry fulminate. Now take a board 10" wide and 20" long, ⅞" thick and with a ¾" bit bore 50 holes through it, (5 rows, 10 to a row). This is the torpedo board. Then take a similar board and with a ½" bit bore the same number of holes in the same position, into it but not quite through. This is the gravel board. A punch will now be required as shown in Fig. 74. a, the nipple being ⅝" diameter and ¾" long. Get some of the best grade tissue paper and cut it in-

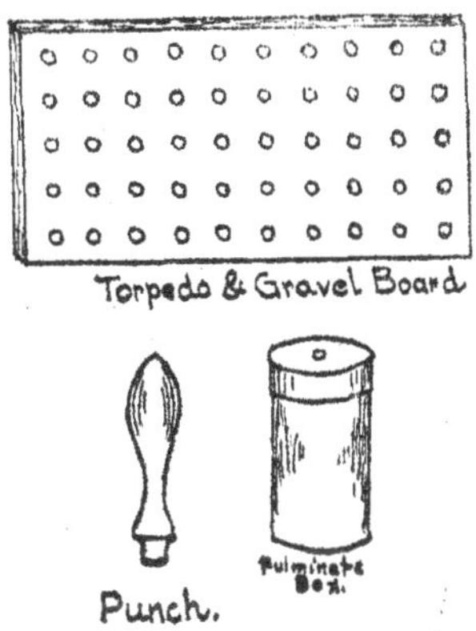

Silver Torpedoes

Fig. 74. a

to pieces two inches square. Take a bunch of them in the left hand. Place one over each hole in the torpedo board, at the same time forcing it in the hole with the puncher so as to make a little bag. When the board is filled with paper dip the gravel board into a box filled with gravel, tilting the board so the surplus will run off and the holes will be just filled. Then reverse the board containing the papers and place it over the board of gravel. Hold tightly together and turn both boards together upside down and the gravel will be emptied into the torpedo board all at once. Remove the now empty gravel board and with the fulminate box shake a little of the powder into each little bag of gravel just as you would shake salt from a salt seller. Only a very little is required. Now dip the tips of the thumb and forefinger into paste and with the finger tips of both hands gather up the edges of the paper, bunch them together and by giving a few twists the torpedo is finished. Care must be taken not to twist too tightly as the torpedo is likely to explode in the fingers.

JAPANESE OR CAP TORPEDOES.

These, while considerably safer than the silver torpedoes to make and handle require to be struck with much more force in order to cause them to explode. First we must proceed to make the caps. (Fig. 75).

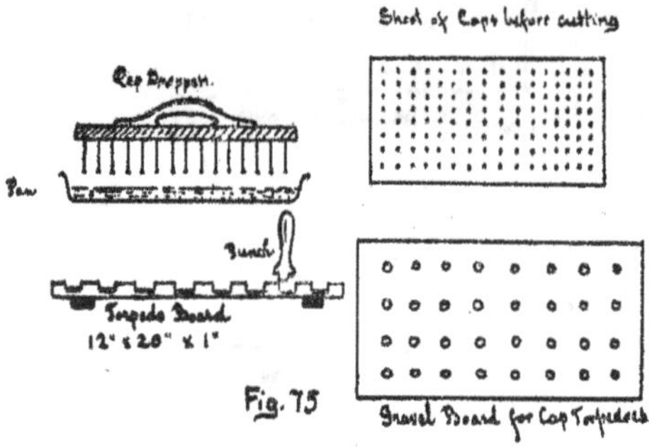

Fig. 75

There are required a pan 2" deep, 8" wide and 10" long, a number of pieces of blanket 12" square which must be well dampened before using, and a cap dropper made by driving 150 8 penny nails, for one inch of their length into a wooden block 7" x 9", 1½" thick and fitted with a handle as shown in illustration. The heads of the nails should be well leveled up so that every one touches when dropper stands on a flat surface. Also cut a lot of pieces of poster paper 6" x 8" and place them in two piles on the work table. They are for dropping the caps into. The cap composition is made as follows:

NUMBER ONE.		NUMBER TWO.
Potassium chlorate	5 ozs.	Amorphous phosphorus
Sulphur	1/4 "	2 ozs.
Chalk	1/4 "	

Sift separately the ingredients of No. 1, mix thoroughly and moisten in a bowl with water to the consistency of porridge. In another bowl moisten the two ounces Amorphous Phosphorus, to the same consistency. Then stir the phosphorus into the bowl containing the other ingredients, with a spoon. When thoroughly mixed pour into the pan previously mentioned.

Take the dropper by the handle and dip it into the pan of composition, remove it and print it lightly onto the top sheet of one of the piles of poster. With a wide brush paste the top sheet on the other pile, all over on one side with thin paste to which a little dextrine has been added and reverse it onto the sheet that has just had the caps dropped on it. Now remove the finished sheet of caps to one of the damp blankets and repeat operation, placing a piece of blanket between each sheet of caps until all the composition is used up after which place a light board on top of the pile of alternate caps and blankets, and on this, a weight allowing it to remain for about an hour, after which remove the blankets and place the pile of caps in a tight box where they cannot become dry.

Now fill the torpedo board as before directed only using a somewhat smaller gravel board. Take out a few sheets of caps and with a long pair of scissors cut between each row of caps each way so as to separate them. Place one squarely on top of the gravel in each torpedo and taking a handful of gravel, drop a little on top of each cap. They are now ready to be twisted as described for silver torpedoes. When a number have been finished they should be packed in sawdust or rice shells and removed from the workroom. Too many should not be allowed to accumulate in a pile, for when dry, the explosion of one will sometimes fire the entire lot and the flying stones often cause serious injury. Be sure never to allow the caps to become dry while making the torpedoes, or in the storage box.

In making caps, when a batch has been completed, be very careful to wipe up thoroughly every drop or speck that may be spilled, and wash well the pan and dropper etc. as well as the table, shears and brush used and throw away the washings where they will run off.

Japanese torpedoes do not keep much over a year as the phosphorus decomposes and after awhile disappears entirely from the cap. Following are the principal sizes and packings of torpedoes:

SILVER TORPEDOES.

Name	Size paper	Holes torpedo board	No. in box	Boxes in case
Electric	1⅝" sq.	½"	25	50
Giant	3" sq.	1"	10	50

JAPANESE TORPEDOES.

Am. Extras	1⅝"	½"	25	100
Japanese	3"	1"	5	200
Japanese	3"	1"	10	100
Japanese	3"	1"	25	40
Cat Scat	5½"	1½"	10	40
Cannon	7"	2"	10	25

RAILROAD TORPEDOES.

Fig. 74b.

These consist of a 1 oz. tin ointment can containing a mixture similar to that used in paper caps. A strip of lead is soldered to bottom of box so it can be easily attached to rail by bending strip around top of rail, and fires when engine runs over it. (Fig. 74 b).

PAPER CAPS.

for toy pistols.

These are made similar to those described under Japanese Torpedoes with such variations of detail as are necessary to their special requirements. They are of course punched out by machinery, a sheet at a time.

WHISTLING FIREWORKS

The peculiar property of potassium picrate to whistle while burning has been known for a long time and has been made use of for producing the amusing whistling fireworks. To make this article:

Disolve 1 lb. picric acid in the least possible quantity of boiling water, in a porcelain receptical; add ¼ lb. potassium carbonate, little by little, stirring continuously. Then add 1 lb. powdered saltpeter. Stir thoroughly; allow to stand for one hour and remove to a heavy piece of filter paper placed in a glass funnel where it can drain. When dry crush to fine powder with a wooden roller. Very small quantities should be handled at a time as an explosion will cause disastrous results. The dry powder may be rammed into tubes from ¼" to ¾" diameter and will produce the whistling sound when burned. Bamboo tubes are most effective.

Owing to the ease with which potass; picrate detonates whistles cannot be used in shells but small tubes ¼" diameter and 2½" long when charged with the above composition may be place in the heads of rockets or fastened to the outside and arranged to burn as the rocket is ascending. Attached to wheels they are quite amusing, but the most effective use for them is when a series of six or eight ranging in size from ¼" to ¾" diameter are set side by side like a Pandaen Pipe and burned simultaneously.

SON OF A GUN.

(Spit Devil—Devil on the Walk).

This amusing little piece of fireworks consist of a

disc about one inch or more in diameter which, when scratched on the pavement gives off a continuous series of little explosions, burning from one half to three quarters of a minute. On account of somewhat resembling candy losenges a number of fatal accidents have occured by small children swallowing them. For this reason their sale has been forbidden in some sections. They are made as follows: (Fig. 76)

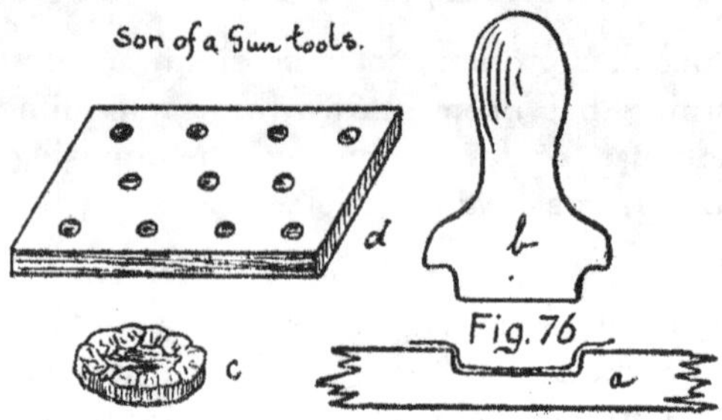

Fig. 76

Secure a number of boards of ⅞" material and bore holes into them ¼" deep and 1¼ diameter somewhat as shown at (a). Turn up a puncher like (b) which will work easily in the holes. Cut some red Foudrinier tissue paper into circular pieces 2½" diameter. Lay them over the holes in board and punch in. Pour into these the composition given below and fold over the edges of the paper as in (c). Permit to set and when hardened they may be removed and thoroughly dried when they will be ready for use.

SON OF A GUN COMPOSITION.

Mix five kilograms of powdered gum Arabic with five litres of water, adding water gradually with constant stirring. Then add 1½ kg. Magnesium carbonate. Place this in a water bath with a thermometer arranged so that temperature can be carefully observed and heat to 50°C. after which add a mixture of one kg. white phosphorus and stir until entirely melted. Continue stirring while cooling to 25°C. after which add a mixture of 2½ kg. red ochre and 3 kg. potass; chlorate and stir until a perfectly smooth product results after which it may be poured into the paper molds as above. Great care must be used to prevent accidents in all mixtures containing phosphorus and chlorate of potass.

PHARAOHS SERPENTS EGGS.

This remarkable substance consists of small pellets of sulphocyanide of mercury which has the remarkable property of swelling 25 to 50 times its original size when lighted, producing a long snake like ash. To prepare, make a concentrated solution of mercuric chloride and add little by little a solution of potassium sulphocyanide, stirring constantly. A greyish precipitate will be formed and when the last drop of sulphocyanide added no longer produces cloudiness permit the mixture to settle. Drain off as much as possible of the clear supernant liquid, remove precipitate to a paper filter placed in a glass funnel and wash slightly. When thoroughly dry reduce to a fine powder. When ready to form the eggs moisten very sparingly with a week solution of gum arabic to which may be added a pinch of saltpeter and form into cones with the appliance shown in Fig. 77.

Fig. 77

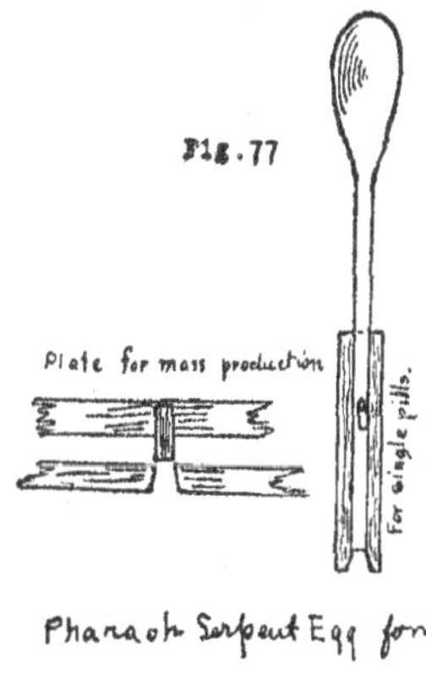

Pharaoh Serpent Egg former.

MAGIC SERPENT, (Black).

This German device produces an immense long black snake, otherwise quite similar to the Pharaohs Serpents but in no ways related chemically.

Naptha pitch	10
Linseed oil	2
Fuming nitric acid	7
Picric acid	3½

Reduce pitch to fine powder; add linseed oil and rub in well; add strongest fuming nicric acid, little at a time. Allow to cool for one hour. Wash several times with water, the last time allowing mass to stand in the water for several hours. Thoroughly dry; powder fine and add picric acid, rubbing it in well. Moisten with gum arabic water and form into pellets about the size of a #4 star.

The naptha pitch can only be obtained in Germany and then with considerable difficulty. A fairly good article may be made by melting together equal parts of Syrian asphaltum and roofing pitch. To the final product add 5% stearine when forming into stars.

SNAKE NESTS.
(Snake in the grass).

Fig. 78.

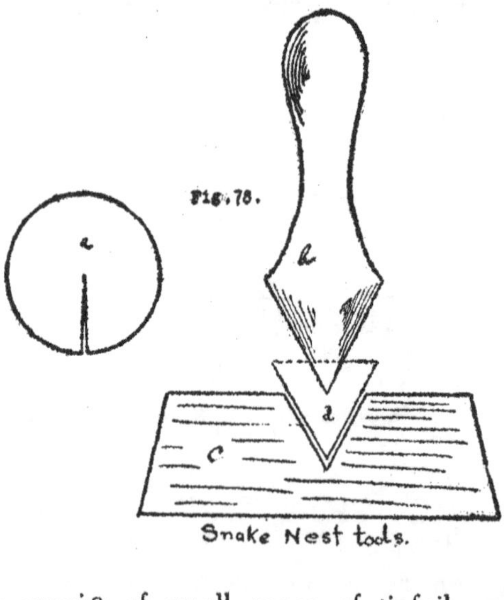

Snake Nest tools.

These consist of small cones of tinfoil containing a preparation which, when ignited, produces a grass green pile of ash from which presently emerges a "Pharaohs serpent".

Cut some tinfoil into circles 1½" diameter. Cut these again from the periphery to the center as shown in Fig. 78. (a). Fold them around the former (b) so as to make little cones and insert into block (c) filling them with the following composition:

Saltpeter	1
Ammonium bichromate	2
Dextrine	1

When quite full up to the edge of the block, press into center a pellet of Pharaoh's Serpent powder. Fold over the edges to the center and remove from block. To use, light at top of cone.

COLORED FLAMES.

These are made by disolving various substances in alcohol. A copper can filled with cotton is impregnated with the alcoholic solution. It is lighted by a tuft of cotton left protruding from the opening. (Fig. 79).

Fig. 79

For Green Flame	use Boric acid
Red "	Strontium or Lithum chloride
Yellow "	Sodium chloride
Blue "	Copper sulfate or Caseium carbonate

Before lighting sprinkle a little of the powdered chemical over the cotton where it projects from the can.

COLORED FIRE STICKS.

These consist of thin wooden sticks similar to applicators used by physicians for applying Iodine etc. to affected parts. They are dipped for half their length into colored fire compositions in a more or less liquid state.

Fig. 80

One method is to melt one part of gum shellac in an, iron pot. Stir in five parts of very finely powdered strontium nitrate. To keep this sufficiently liquid it must be kept quite hot by the use of a steam kettle. This is for red sticks. Another method is to disolve the shellac in alcohol and adding the strontium. The proper consistency of the mixture can be easily regulated by using more or less alcohol as required. When the sticks are dried they are ready for use.

Green is not so successfully made, barium nitrate being substituted for strontium. A little lampblack improves the burning but detracts from the color, especially the green. The sticks are pushed into a grove in the bar as shown, Fig. 80 for drying.

Ruby and Emerald Shower Sticks.

These are much more effective and are made in the same manner as above, using following composition:

Strontium nitrate	6
Coarse aluminum	6
Potass perchlorate	2
Shellac	1

Disolve shellac in alcohol and add other ingredients, previously well mixed. Stir thoroughly to consistency of thick glue and dip sticks previously arranged in holder so they may be placed in drying rack.

For GREEN use:

Aluminum powder coarse	6
Barium chlorate	4
Shellac	¾
Alcohol q. s.	

The Japanese make a similar article of twisted paper but this requires a great deal of practice to learn, almost unattainable by western races.

SPARKLERS.

These are made in a general way like the above but in effect they throw off a shower of beautiful sparks. There are several varieties of sparkling sticks which are sold under this name. The principal one consists of pieces of wire or thin twisted metal, part of which is covered with a composition containing steel filings.

STEEL SPARKLER.

Fine steel filings	12
Fine aluminum powder	1
Potassium perchlorate	6
Dextrine or gum arabic	2
Water q. s.	

The steel must be protected from corosion with parafine. The gum should be made of the consistence of mucilage. Mix the ingredients thoroughly and add gum solution until a mixture is obtained that will adhere to the wires when they are dipped into it. This varies in different sections and with different runs of ingredients. In practice, bunches of wires are dipped at once and slowly withdrawn in a current of warm, dry air which causes the mixture to adhere evenly.

A sparkler of great brilliance and which is very effective may be made as follows: Take 3 lbs. of dextrine and add to same, little at a time, 12 pints of water, stirring continually so as to avoid lumps. Mix intimately 10 lbs. potassium perchlorate with 7 lbs. pyro-aluminum or finely powdered aluminum and add this to the gum water, stirring until a perfectly smooth mixture is obtained. Wood sticks may now be dipped into it to the desired depth while it is contained in a deep vessel, and placed in a suitable rack for drying. It may be necessary to dip the sticks several times dependent on how much composition it is desired to have on them. In this case they should be dried with the composition end up, the first time so that not to much composition accumulates on the end beyond the stick.

WATER FIREWORKS.

These consist mainly of five or six varieties as follows: (Fig. 81).

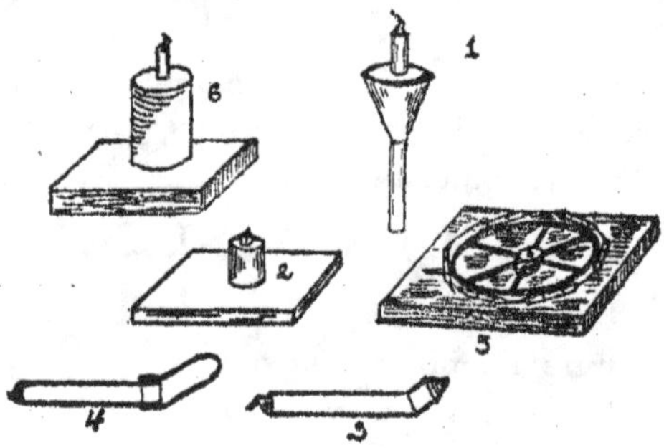

Fig. 81

No. 1 Floating gerb or roman candle. A cone shaped piece of light wood is bored with hole of suitable size to take the gerb or candle as shown. In order to secure an upright position with roman candles it is sometimes necessary to place a charge of iron filings in the bottom of case.

No. 2. Floating Tableau Lights are merely a colored pot of suitable size placed on a board.

No. 3. Diving Devils. A sharp gerb is fitted with a hollow head set at an angle with the case. Careful adjustment must be made so as to insure floating of the gerb which will cause it to dive into and come up out of the water properly. This is perhaps the most amusing piece of water fireworks as well as the one calling for the most careful work. The tip of float must be weighted so as to cause it to dive and yet. be bouyant enough to make it rise again.

No. 4. Fish are made similar to the Diving Devil except that not so much adjustment is necessary as they only run around on top of the water.

No. 5. Water Wheels are an ordinary Vertical Wheel set on a board float as shown.

The Fish and Devils should be heavily coated with parafine when finished; even the nosing of the match should be protected in this manner and water proof fuse, properly primed, used for lighting.

Water fireworks are only practical on quiet ponds and small lakes and are usually fired from a skiff. Great care should be used to protect the supply in boat from sparks of those burning, in order to prevent accidents to the operator. #6 is a Floating Mine.

SMOKE and SPARK POTS.

These are used mostly in the large spectacular pyrotechnical pictures featuring volcanos etc. such as "The last days of Pompeii" "Burning of Rome" etc. They consist of short mine cases about 4" to 6" in diameter and 6" to 12" long. The compositions are

SPARK POT.

Meal powder	2
Fine charcoal	1
Sawdust	1

SMOKE SCREENS.

While these are really no part of the pyrotechnical art they do come under the heading of Military Pyrotechnics. They consist of a fine stream of Titanium tetrachloride which is sprayed from an airplane at a suitable height and which in falling produces the dense smoke intended to screen what is behind it. The liquid is projected backward from the plane at the same rate as the planes movement through the air so that the droplets fall perpendicularly.

SMOKE.

This branch of pyrotechny seems to have been somewhat overlooked though its possibilities for daylight entertainment to supplement night displays would appear to open an interesting field for those with the necessary imagination to develope it.

There are as many colors and tints of smoke as there are of flames and aerial combinations produce some really beautiful effects

The simplest form of smoke used in pyrotechny is the Smoke Pot used in the spectacles such as "Last Days of Pompei" where it is desired to give the impression of destruction by fire. A basic formular, subject to variation is:

Saltpeter	4
Lampblack	1
Charcoal	1
Realgar	1
Rosin	1

A starting fire as given below is also necessary.

A smoke pot developed by our War Department for making smoke screens consists of case 3½" diameter 6" high with a 1" opening in the top. The composition consists of:

Saltpeter	12
Pitch	8
Borax	2½
Chalk	1¼
Sand	1
Sulphur	1

The pitch is melted and while still hot the other ingredients, previously well mixed, are stirred in. Before the mixture cools and hardens it is pressed into case. As it is difficult to ignite, a starting fire of

Saltpeter	6
Sulphur	1¼
Antimony sulfid.	1
Meal powder	1

is used. About ¼" of this is placed on top of other composition before putting the top on case. This may consist of a wooden disc with hole, fastened in place with small nails or a tin can may be used as container.

SMOKE SHELLS.

(Smoke Clouds)

These are usually made by filling the shell case with a finely divided powder of the desired color cloud to be obtained. To the end of the shell fuse is attached a small bag of gunpowder which should be located as near the center of the shell as possible. This, when exploding serves to scatter the colored matter and produce the cloud. The arrangement of the fuse may be used as shown in Fig. 67.

For RED	use	American vermillion powder
BLUE		Ultramarine powder
GREEN		Paris green
YELLOW		Chrome yellow
WHITE		Chalk
BLACK		Ivory black

SMOKE POTS or SMOKE CASES.

The really beautiful effects, however, are produced by the burning of mixtures which produce a dense smoke of the desired color. For this purpose a paper tube 1 inch inside diameter and 4 inch long is desirable, into which are bored 4 or 5 holes ¼" diameter on a spiral line, at intervals as shown in Fig. 86. Both ends of the case may be closed with clay or wooden plugs. Do not pack smoke compositions. Ram very lightly.

WHITE SMOKE.

Potassium chlorate	3
Lactose	1
Sal ammoniac, finely powd.	3

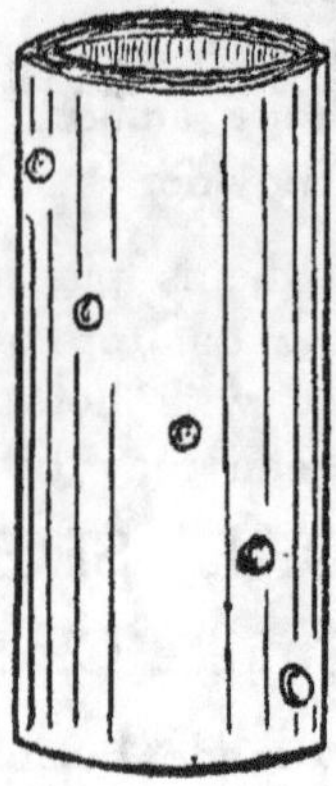

As this smoke is not injurious to the lungs it may be used in doors for theatrical work. Rub the potass: chlorate and lactose together thoroughly, then add sal ammoniac, finely powdered, mixing only once more. Ram lightly. Another very good white smoke may be made as follows:

Sulphur flowers	16
Saltpeter	12
Fine charcoal	1

This is for outdoor use only. Use a little white star composition for starting fire.

BLACK SMOKE.

Hexichloroethane	24
Alpha naptha	6
Anthracene	2
Aluminum powder	4
Roman candle composition	6

Use white star composition for starting, if found necessary.

RED SMOKE. (bright)

Potassium chlorate	1
Lactose	1
Paranitraniline red	3

RED SMOKE. (dark)

Potassium chlorate	7
Lactose	5
Auramine	2
Chrysoidin	6

YELLOW SMOKE. (canary)

Potassium chlorate	1
Lactose	1
Paranitraniline yellow	2

Use red star composition for starting either of the above three.

YELLOW SMOKE. (olive)

Saltpeter	1
Red arsenic	1
Sulphur	1
Antimony sulphide	1
Meal powder	1

No starting fire necessary.

GREEN SMOKE. *

Potassium chlorate	6
Lactose	5
Auramine	3
Indigo (synthetic)	5

BLUE SMOKE.

Potassium chlorate	7
Lactose	5
Indigo (synthetic)	8

Starting fire if needed.

Almost any pastel shade desired may be produced by combining the above formulas containing aniline dyes.

PART IV.

EXHIBITION FIREWORKS.

SNAKE and BUTTERFLY

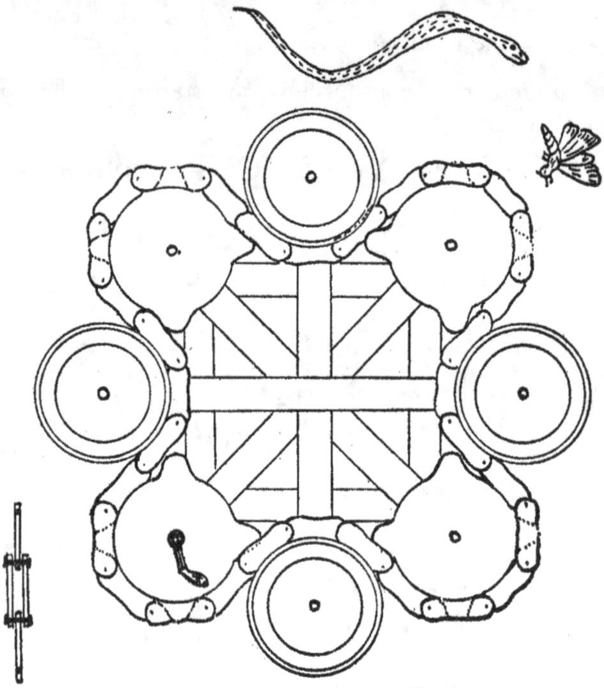

Fig. 82

This ingenious piece of fireworks is believed to have been devised by the Brock Fireworks Co. of Surrey, England, and created quite a sensation at the Crystal Palace when first shown about forty years ago. It consists of a snake squirming around in the air after a butterfly which manages however to evade it.

The framework consists of an endless chain of wooden links 4" x 8" bolted together and running on four sprockets and four idlers of a suitable size as shown in Fig. 82. When mounted, a crank is attached to one of the sprockets by which the whole is operated. The snake and butterfly are made of lancework which is attached to the chain.

ROCKET WHEEL

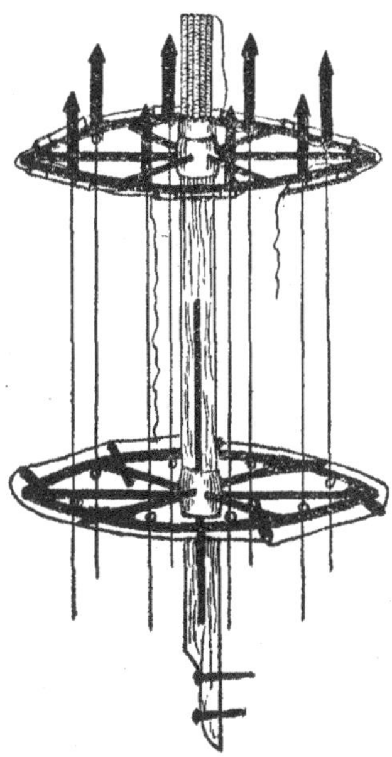

Rocket Wheel.

Fig. 83.

This is a very old, yet always attractive device. It consists of two wheels three feet in diameter, attached to opposite ends of an axle, arranged to revolve horizontally on a spindle as shown. The rockets pass through screw eyes along the rim of the wheel, and are matched to fire at intervals as the wheel revolves by being connected to successive drivers. On the top is a battery of roman candles. The top wheel is fitted with ordinary drivers containing steel filings and matched to burn two at a time, one each on opposite sides. The lower wheel is fitted with aluminum gerbs all to burn at once with the last two drivers of the top wheel. The latter are set at an angle with the axis of rotation so as to give a wider spread of fire. The battery of candles starts with second pair of drivers of top wheel. (Fig. 83).

REVOLVING GLOBE.

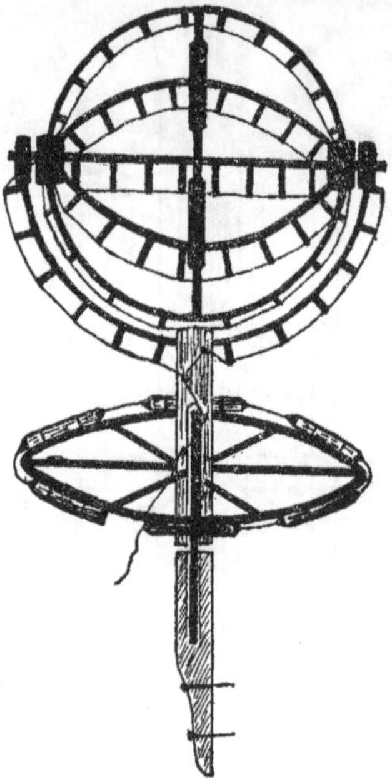

Revolving Globe.

Fig. 84.

This simple yet baffling and always interesting device is constructed as shown in Fig. 84. The frame may be secured in different sizes, all ready for lances etc. from manufacturers of fireworks wheels in North Weare, N. H. or it may be constructed by the pyrotechnist himself according to suggestions given in sketch.

When the piece is burning, the globe appears to be revolving first in one, then in the other direction in a most amusing manner.

APPENDIX.

CHINESE FIRE CRACKERS.

As far as is known to the writer there has never been given in English a detailed description of this interesting little article of pyrotechnics of which there are undoubtedly more made than of any other piece of fireworks. The ingenuity of the Chinese in its production in the unbelievable large quantities that they are made is only equaled by the many other unusual things done by this most patient and painstaking race.

The yearly imports to this country of Chinese firecrackers amounts to three million dollars which divided among the various sizes, would amount conservatively to eight billion crackers.

The tubes or firecracker cases cases are 1¾" long, ¼" outside diameter and have a bore of 5/32". They are rolled of a grade of paper unknown in this country; perhaps the lowest grade of paper made, unsized and quite irregular in character; a sort of coarse blotting paper. A small amount of gum water or rice paste is used as a binder and the case is finished with one turn of very thin red, green or yellow paper. They are rolled in lengths of one to two feet and then cut to the required size.

Now a block is prepared for gathering about 1000 of these tubes into a hexagon shaped bundle, as follows. A piece of hard wood about 1" thick and cut into a hexagon, each side of which is 5" wide is provided with

pointed wood or metal pins ¾" long and 5/32" diameter set into the wood base so that the above amount projects, and exactly ¼" apart. They are also arranged in a hexagon with sides 4" wide. A tube is now slipped over each pin until the entire block is filled, having previously provided a wood frame the same size as the outside of the block one half inch thick and having an inside diameter slightly greater than the assembled tubes so as to be able to slip snugly around them. This is slipped up and down a few times to shape the bundle nicely and a string tied around it to further secure same.

A piece of white paper is now pasted over the top of the bundle. When dry it is removed from the form and a piece of paper pasted on the other side when it is dried again. The under side is moistened at the edges and the surplus paper neatly rubbed off. When again dry the upper side is moistened all over and the paper over the top of each cracker is pierced with a punch or round pointed stick so that they may be charged with the necessary powder and clay. Some operators hold several sticks between their fingers at one time so as to be able to punch several holes simultaneously.

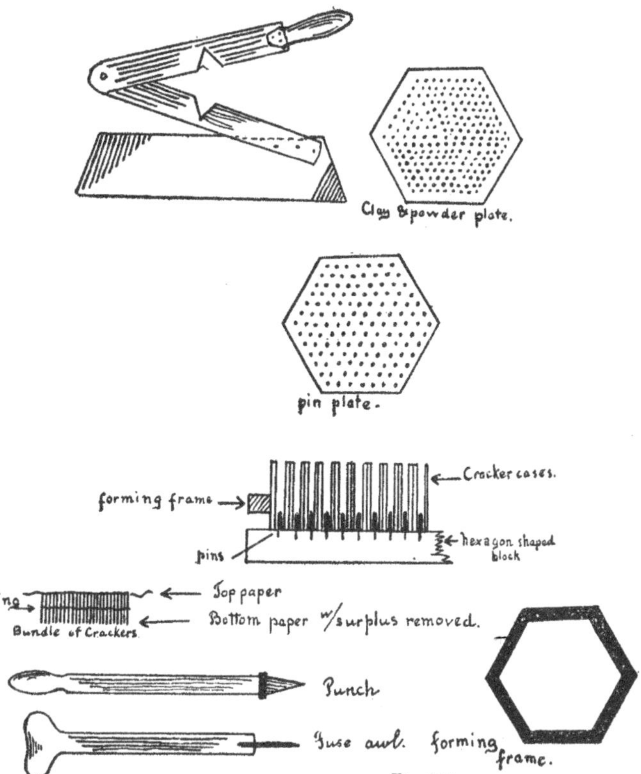

Fig. 85.

A wooden board about one inch wider all around than the bundle of crackers and ¼" thick with ⅛" holes bored through it, corresponding exactly in position with the crackers in the bundle, is now laid on a smooth board, covered with finely powdered clay which is pressed into the holes in it, with the hand, until it is firm enough not to fall out when the piece is lifted. The surplus is brushed off and it is placed over the bundle of crackers so that the clay filled holes are exactly over the openings in the tubes. A slight blow is usually sufficient to to cause the clay to fall into the crackers. Any not falling out is pushed out with a stick. The bundle is jarred slightly against the table to make the clay settle. A similar operation is now performed with a thicker board containing slightly larger holes containing the powder charge after which the clay board is used once more as described above.

The top layer of paper is now moistened so that it it may be entirely removed and the clay which has become slightly moistened as well, is gently pushed down with a suitable rammer. It is then dried in the sun. The bottom end is now carefully dipped into water, turned bottom up and the paper removed from this side also, the clay pushed down and pierced with an awl for the purpose of inserting the match or fuse. This is however not done until the crackers have been again dried in the sun. After the fuses are inserted the ends of the crackers are pinched around it, about ⅛" from the end, by a crimper or two blunt knives hinged together at one end and having a V shaped notch cut out of the center of each blade, so that when two notches approach from opposite sides they pinch the cracker together and cause the fuse to be held in place. When they are now finally dried for the last time they are platted together so as to form the packs of commerce. The platting and wrapping of the packs is such a dextrous performance that it is useless to try to describe it as it is only acquired by many years of succeeding generations doing the same thing.

The following formulas are in use for making the composition used in Chinese crackers and flash crackers:

CHINESE FIRE CRACKERS.

Saltpeter	50	45
Sulphur	25	18
Charcoal	25	25
Chlorate potassium		8
Sand		4

FLASH CRACKERS.

Saltpeter	50		
Sulphur	30	25	30
Aluminum powder fine	20	25	40
Chlorate potassium		50	30

A very important as well as extremely difficult part of the Chinese cracker is to make the fuse. Very tender and skilled fingers are required to produce this insignificant looking yet most requisite adjunct. A thin strip of the finest Chinese tissue paper, about ¾" wide and 14" long is laid on a smooth damp board; a little stream of powder is poured down its center from a hollow bamboo stick and with the tips of soft skinned fingers which seem to have an attraction for the paper and placed against the right hand lower corner, a rolling motion in the general direction of the upper left hand corner causes the paper to roll up into a twine like fuse. The slightest touch of paste secures the end and prevents unrolling. When dry it is cut into the required lengths and is ready for use.

NOTE. The information upon which the foregoing article was written has been supplied by Mr. Ip Lan Chuen, manager of the Kwong Man Loong Fireworks Co. of Hong Kong, China.

INDEX